Recent Advances in Metal Powder Based Additive Manufacturing

Recent Advances in Metal Powder Based Additive Manufacturing

Guest Editors

Hong Wu

Yingtao Tian

Alberto Orozco Caballero

MDPI

Basel • Beijing • Wuhan • Barcelona • Belgrade • Novi Sad • Cluj • Manchester

Guest Editors
Hong Wu
State Key Laboratory of Powder Metallurgy
Central South University
Changsha
China

Yingtao Tian
School of Engineering
Lancaster University
Lancaster
UK

Alberto Orozco Caballero
Department of Mechanical Engineering
Universidad Politécnica de Madrid
Ronda de Valencia
Spain

Editorial Office
MDPI AG
Grosspeteranlage 5
4052 Basel, Switzerland

This is a reprint of the Special Issue, published open access by the journal *Materials* (ISSN 1996-1944), freely accessible at: https://www.mdpi.com/journal/materials/special_issues/metal_powder_AM.

For citation purposes, cite each article independently as indicated on the article page online and as indicated below:

Lastname, A.A.; Lastname, B.B. Article Title. *Journal Name* **Year**, *Volume Number*, Page Range.

ISBN 978-3-7258-4453-1 (Hbk)
ISBN 978-3-7258-4454-8 (PDF)
https://doi.org/10.3390/books978-3-7258-4454-8

Contents

Preface

The rapid evolution of metal powder-based additive manufacturing (AM) presents unprecedented opportunities for materials innovation and sustainable manufacturing. As Guest Editors of this Special Issue Reprint, we are privileged to present this curated collection of pioneering research that addresses both the scientific complexities and engineering challenges inherent to this transformative field. Our motivation stemmed from the urgent need to consolidate emerging knowledge on AM-compatible material systems while fostering interdisciplinary dialog between process engineers, metallurgists, and industrial practitioners.

This Reprint comprises several rigorously peer-reviewed contributions that collectively advance our understanding of microstructure control, alloy design strategies, and process–property relationships in powder-based AM. The works span fundamental investigations into thermal gradient effects during layer-wise fabrication, computational modeling of defect evolution, and innovative approaches for multi-material integration. Particular attention is given to overcoming the limitations of conventional materials in AM environments, with several studies proposing novel high-entropy alloys and refractory metal processing techniques.

We extend our deepest gratitude to the international cohort of authors—researchers from academia, national laboratories, and industry—whose expertise illuminates diverse aspects of AM development. Their contributions not only validate new characterization methodologies but also establish practical frameworks for quality assurance in critical applications. The Reprint is particularly relevant for materials scientists working on phase-transformation mechanisms, engineers optimizing post-processing protocols, and policymakers shaping standards for AM industrialization.

We also appreciate the Editorial team at *Materials* for their professional support throughout the publication process. To early-career researchers entering this dynamic field, we hope this Reprint serves as both an educational resource and an inspiration to tackle unresolved challenges in process scalability and material recyclability.

Hong Wu, Yingtao Tian, and Alberto Orozco Caballero
Guest Editors

materials

Editorial

Recent Advances in Metal Powder-Based Additive Manufacturing

Hong Wu [1,*], Yaojia Ren [1], Yingtao Tian [2] and Alberto Orozco Caballero [3]

1 State Key Laboratory of Powder Metallurgy, Central South University, Changsha 410083, China
2 Department of Engineering, Lancaster University, Lancaster LA1 4YW, UK
3 Department of Mechanical Engineering, Chemistry and Industrial Design, Universidad Politécnica de Madrid, Ronda de Valencia, 3, 28012 Madrid, Spain
* Correspondence: hwucsu@csu.edu.cn

Over the past two decades, laser additive manufacturing technology has evolved rapidly and has been applied in many industrial sectors. However, traditional manufacturing methods such as casting, forging, rolling, and welding are still prevalent around the world [1]. To meet the demand for lightweight and personalized applications in aerospace and biomedical fields, the use of laser powder bed fusion (LPBF) is proliferating and is expected to maintain a high growth rate in the next ten years.

Significant technical advances have been made in the field of metal powder-based additive manufacturing, and a range of high-performance materials have been successfully developed [2,3]. However, the poor printability and unacceptable metallurgical defects are still the primary concerns that limit the adoption of most alloys in the laser-powder bed-fusion process [4]. Therefore, it is essential to re-design the alloy compositions and make them suitable for LPBF which requires a comprehensive understanding of the impact of complex thermal cycles to the microstructure and properties of the materials. In addition, LPBF offers a superior work-hardening effect to the materials, which gives them a chance to overcome the classical strength-ductility trade-off [5]. This also includes the potential effects and consequences of post-treatment on the microstructure and properties. High internal stresses in as-built specimens are inevitable due to the extremely high cooling rates. This may lead to premature material failure when long service periods are required. Furthermore, understanding the hierarchical heterostructure evolution and the corresponding property changes during post-treatment is of great significance for the development of high-strength materials.

The current Special Issue entitled "Recent Advances in Metal Powder-Based Additive Manufacturing" collects the recent research outcomes addressing the main challenges and aiming to provide possible solutions that may revolutionize the metal powder-based additive manufacturing technology and its applications. It is believed that this Special Issue will provide an innovation platform for researchers in this area to communicate and disseminate their most recent ideas and achievements which will facilitate and support scientists and engineers to continuously make contributions to the field.

Conflicts of Interest: The authors declare no conflict of interest.

References

1. Abd-Elaziem, W.; Elkatatny, S.; Abd-Elaziem, A.-E.; Khedr, M.; El-Baky, M.A.A.; Hassan, M.A.; Abu-Okail, M.; Mohammed, M.; Järvenpää, A.; Allam, T.; et al. On the current research progress of metallic materials fabricated by laser powder bed fusion process: A review. *J. Mater. Res. Technol.* **2022**, *20*, 681–707. [CrossRef]
2. Wu, H.; Ren, Y.; Ren, J.; Liang, L.; Li, R.; Fang, Q.; Cai, A.; Shan, Q.; Tian, Y.; Baker, I. Selective laser melted AlSi10Mg alloy under melting mode transition: Microstructure evolution, nanomechanical behaviors and tensile properties. *J. Alloy. Compd.* **2021**, *873*, 159823. [CrossRef]
3. Ren, J.; Zhang, Y.; Zhao, D.; Chen, Y.; Guan, S.; Liu, Y.; Liu, L.; Peng, S.; Kong, F.; Poplawsky, J.D.; et al. Strong yet ductile nanolamellar high-entropy alloys by additive manufacturing. *Nature* **2022**, *608*, 62–68. [CrossRef] [PubMed]
4. Wang, H.; Chen, L.; Dovgyy, B.; Xu, W.; Sha, A.; Li, X.; Tang, H.; Liu, Y.; Wu, H.; Pham, M.-S. Micro-cracking, microstructure and mechanical properties of Hastelloy-X alloy printed by laser powder bed fusion: As-built, annealed and hot-isostatic pressed. *Addit. Manuf.* **2021**, *873*, 101853. [CrossRef]
5. Wang, Y.M.; Voisin, T.; McKeown, J.T.; Ye, J.; Calta, N.P.; Li, Z.; Zeng, Z.; Zhang, Y.; Chen, W.; Roehling, T.T.; et al. Additively manufactured hierarchical stainless steels with high strength and ductility. *Nat. Mater.* **2018**, *17*, 63–71. [CrossRef] [PubMed]

materials

MDPI

Article

Microstructure and Mechanical Properties of As-Built Ti-6Al-4V and Ti-6Al-7Nb Alloys Produced by Selective Laser Melting Technology

Dorota Laskowska *, Błażej Bałasz and Wojciech Zawadka

Faculty of Mechanical Engineering and Energy, Koszalin University of Technology, Śniadeckich 2, 75-453 Koszalin, Poland; blazej.balasz@tu.koszalin.pl (B.B.); wojciech.zawadka@tu.koszalin.pl (W.Z.)
* Correspondence: dorota.laskowska@tu.koszalin.pl

Abstract: Additive manufacturing from metal powders using selective laser melting technology is gaining increasing interest in various industries. The purpose of this study was to determine the effect of changes in process parameter values on the relative density, microstructure and mechanical properties of Ti-6Al-4V and Ti-6Al-7Nb alloy samples. The experiment was conducted in response to a noticeable gap in the research on the manufacturability of the Ti-6Al-7Nb alloy in SLM technology. This topic is significant given the growing interest in this alloy for biomedical applications. The results of this study indicate that by properly selecting the volumetric energy density (VED), the relative density of the material produced and the surface roughness of the components can be effectively influenced. Microstructural analyses revealed similar patterns in both alloys manufactured under similar conditions, characterized by columnar β phase grains with needle-like α′ phases. Increasing the VED increased the tensile strength of the fabricated Ti-6Al-4V alloy components, while the opposite effect was observed for components fabricated from Ti-6Al-7Nb alloy. At the same time, Ti-6Al-7Nb alloy parts featured higher elongation values, which is desirable from the perspective of biomedical applications.

Keywords: additive manufacturing; selective laser melting; Ti-6Al-4V; Ti-6Al-7Nb; relative density; microstructure; mechanical properties; biomedical applications

1. Introduction

High strength-to-weight ratios characterize titanium and its alloys for a vast range of operating temperature variations and high corrosion resistance in many chemical environments [1–3]. These properties create many opportunities for applications of components manufactured from titanium-based alloys, including aerospace, automotive, chemical and, most importantly, biomedical engineering. Ti-6Al-4V is the most widely used alloy for long-lasting and load-bearing bone implants in biomedical applications. Many studies highlight its high biocompatibility, although this is increasingly questioned due to the presence of vanadium [4,5]. To solve this issue, vanadium-free titanium alloys, like Ti-6Al-7Nb, were developed.

Titanium and its alloys belong to a group of materials defined as hard to machine. The challenges associated with traditional machining methods, such as milling, casting, forging or rolling, increase production costs and make manufactured parts more expensive compared to those from Fe-based alloys or Co-Cr-Mo alloys. These problems have been partially solved by additive manufacturing.

The American Society for Testing Materials (ASTM) has introduced a definition of additive manufacturing (AM) as the process of joining successive layers of material (e.g., powder) based on a 3D model of production parts [6,7]. An important advantage of additive manufacturing, especially in the case of biomedical applications, is customization. AM allows for the production of medical implants tailored to specific patient requirements [8].

Although Ti alloys show a lower elastic modulus compared to other implant materials (e.g., Co-Cr-Mo), it is higher than the elastic modulus for the bone tissue being replaced, which can lead to the so-called stress-shielding phenomenon [9,10]. The characteristics of AM processes make it possible to fabricate components with complex geometry and spatial porosity (designed based on TPMS topology) and to define the directionality of mechanical properties, designing so-called architectural materials [11–14]. As studies show, this approach allows us to reduce the stiffness of the fabricated implant, thus reducing the risk of stress shielding [15].

Additive manufacturing has drawbacks that limit its application in large-scale and mass production. The most significant issues include: a reduction in mechanical properties due to internal porosity [16]; the anisotropy of mechanical properties of the produced components [17,18]; the dependence of precision and quality on various, often interrelated or mutually exclusive factors, such as the properties of the construction material or process parameters, and the necessity of post-processing to improve dimensional and shape accuracy, mechanical properties, or surface quality [19].

A widely used process for the additive manufacturing of titanium and its alloys is selective laser melting (SLM). In this technology, a laser is used as a source of thermal energy to melt and fuse a specific volume of powder [20]. The process is carried out in a protective gas atmosphere to prevent the oxidation of the molten material. The laser beam induces only part of the generation of the heat energy which is absorbed by the powder grains, while the rest of the laser beam is reflected without affecting the melting process. The local generation of thermal energy results in the formation of a molten metal pool, in which phenomena such as conduction melting [21,22], keyhole melting [23], Marangoni convection [24–26], alloying element segregation [19], evaporation and spattering [27] take place. The occurrence and intensity of the above-mentioned phenomena depend mainly on the volumetric energy density, but also on the conditions in the working chamber (i.e., the residual oxygen content, the temperature of the working platform, or the temperature inside the chamber).

The state of the art on the effect of microstructure on the mechanical properties of additively manufactured Ti-4Al-6V alloy components using SLM technology is readily available thanks to the research results published in numerous publications [28–31]. However, research on the Ti-6Al-7Nb alloy has been conducted to a much lesser extent.

Chlebus et al. [32] found that the Ti-6Al-7Nb alloy produced by SLM technology has a microstructure of columnar grains of primary β phase with long, thin α' martensite plates. This results in higher tensile and compressive strength but lower ductility compared to the alloy produced by conventional methods. In addition, it was pointed out that the microstructure of the alloy produced by SLM technology depends not only on the process parameters but also on the spatial orientation of the manufactured object relative to the build platform. Similar conclusions were reached by a team led by Xu et al. [33], who investigated the effect of microstructure on the properties of Ti-6Al-7Nb and Ti-6Al-4V alloys. In their research, the team added a heat treatment process to the Ti-6Al-7Nb alloy. This reduced the tensile strength and hardness while increasing the elongation of the alloy. This would undoubtedly have a positive impact on the use of Ti-6Al-7Nb alloy in the production of medical implants. The referenced studies on the production of the Ti-6Al-7Nb alloy did not account for the variability in manufacturing parameters—the alloy was produced using a single strategy. From the perspective of applying SLM technology to manufacture Ti-6Al-7Nb components for biomedical applications, it seems appropriate to adopt a more comprehensive approach.

Therefore, the purpose of this study, the results of which are presented in this article, was to determine the effect of process parameters (scanning speed and laser power) and thus the variation of the volumetric energy density on the relative density, microstructure, and mechanical properties of samples made of the Ti-6Al-4V and Ti-6Al-7Nb alloys. A comparison was made between the quality and properties of samples produced using the same manufacturing strategies for the materials under investigation. The aim was to identify the optimal (within the studied range) manufacturing strategy. The primary criterion for selecting the values of the process parameters was to maximize the relative density while minimizing the surface roughness, under the assumption that improvements in mechanical properties can be achieved by subsequent heat treatment. The described studies serve as a starting point (preliminary research) for the subsequent research stages, which will involve developing guidelines for heat treatment of the Ti-6Al-7Nb alloy to improve its mechanical properties, as well as guidelines for post-processing to reduce surface roughness.

2. Materials and Methods

2.1. Powders' Characterization

The samples were fabricated from commercial Ti-6Al-4V powder (3D Systems, Rock Hill, SC, USA) and Ti-6Al-7Nb powder (SLM Solution Group Ag, Lubeka, Germany). The chemical composition of the powders is shown in Table 1. Scanning electron microscope images (Figures 1A and 2A) show spherical grain morphology with satellite characteristics for powders produced by gas atomization technology. The particle size distribution of the powder (Figures 1B and 2B) was determined using an ANALYSETTE 22 MicroTec Plus laser particle size analyzer (Fritsch GmbH, Amberg, Germany) and presented according to PN-ISO 9276-1 [34].

Table 1. Chemical compositions of Ti-6Al-4V and Ti-6Al-7Nb powders (wt. %).

Powder	Ti	Al	V	Nb	Fe	O	C	N	H
Ti-6Al-4V	Balance	6.00	4.00	-	≤0.25	≤0.13	≤0.08	≤0.03	≤0.012
Ti-6Al-7Nb	Balance	6.05	-	7.1	0.15	0.08	0.015	0.016	0.001

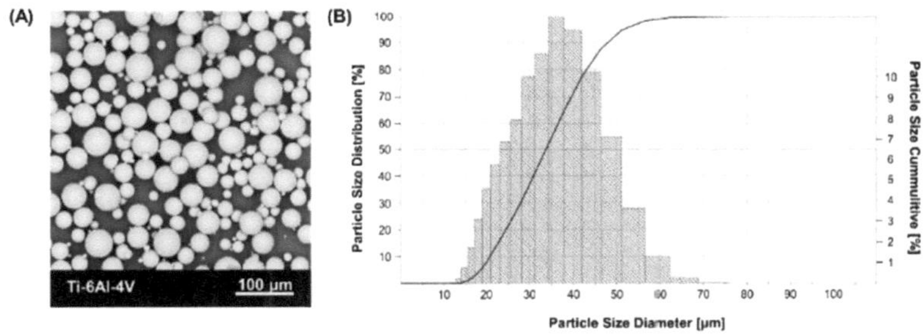

Figure 1. Ti-6Al-4V powder characterization: (**A**) morphology (SEM), (**B**) particle size distribution (PSD).

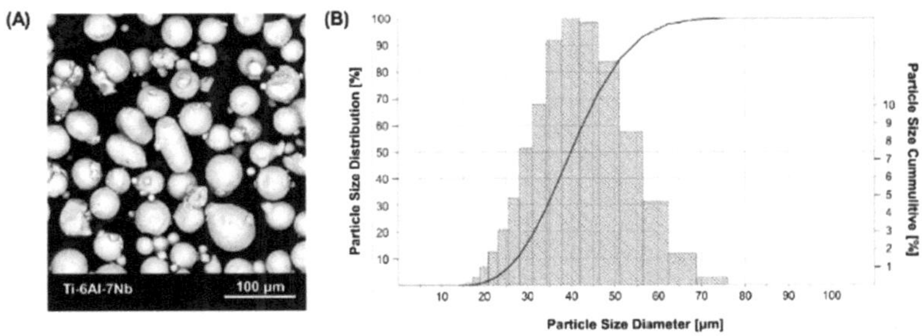

Figure 2. Ti-6Al-7Nb powder characterization: (**A**) morphology (SEM), (**B**) particle size distribution (PSD).

2.2. Samples Fabrication

The samples were fabricated on an ORLAS CREATOR® selective laser melting system (O. R. Lasertechnologie GmbH, Dieburg, Germany) with a Ytterbium fiber laser, with beam spot size 40 μm, a maximum power of 250 W and a wavelength of 1070 nm. The protective atmosphere of the working chamber was provided by the use of argon gas, which allowed the process to be performed at residual oxygen levels below 0.1%.

Volumetric energy density (E_V) is regarded as a key metric for evaluating the complex parameters involved in selective laser melting technology, and is defined by the following equation [35,36]:

$$E_V = P/(v \cdot h \cdot t), \quad (1)$$

where P—laser power [W], v—scan speed [mm/s], h—hatch distance [mm], and t—layer thickness [mm]. In this work, the combination of values of individual parameters was chosen so that the volumetric energy density was in the range of 55–70 J/mm^3. In this way, 9 manufacturing strategies were developed (Table 2).

Table 2. Manufacturing strategies for Ti-6Al-4V and Ti-6Al-7Nb.

Manufacturing Strategy Symbol	Parameters				
	P [W]	v [mm/s]	h [mm]	t [mm]	E_v [J/mm^3]
S0	216	1200	0.1	0.03	60
S1		1430			50
S2		1300			55
S3		1100			65
S4		1030			70
S5	180	1200			50
S6	199				55
S7	233				65
S8	250				70

For each manufacturing strategy, 3 cubic samples (with dimensions of 10 × 10 × 10 mm) and 5 tensile samples (dimensions defined by PN-EN ISO 6892-1:2020-05 [37]) were produced. Post-manufacturing sample preparation included mechanically removing supports and ultrasonically cleaning the samples in distilled water for 10 min.

2.3. Relative Density

The relative density of the manufactured samples was measured using a Mettler Toledo XS105 hydrostatic balance (Mettler Toledo, Columbus, OH, USA). Three measurements were made for each sample, resulting in nine measurements for each manufacturing strategy.

2.4. Surface Morphology

The surface morphology was analyzed using a PHENOM PRO scanning electron microscope (Thermo Fisher Inc., Waltham, MA, USA), with magnification in the range of 160–350,000× and a resolution of ×6 nm. Images of the top surface in the build state were captured for a randomly selected sample from each batch.

2.5. Surface Topography and Roughness

Surface topography was analyzed using the Olympus LEXT OLS4000 confocal microscope (Olympus, Shinjuku, Tokyo, Japan) for the upper surface of each fabricated specimen. The data acquisition area was set to 3 × 3 mm. Surface topography images were analyzed using TalyMap Platinum v7.4. software (Taylor Hobson, Leicester, UK). The data were filtered using a Gaussian filter with a length of 0.8 mm. The features of the additively manufactured surfaces were evaluated using the parameters of arithmetic mean height (Sa) and maximum height (Sz) (according to ISO 25178-2:2021 [38]), which are the most frequently used when assessing this type of surface [39,40].

2.6. Microstructure and Structural Defects

The structural defects analysis was conducted using the optical microscope NIKON MA200 (Nikon, Minato, Tokyo, Japan) on resin-embedded vertical cross-sections of randomly selected samples from each series. The samples were cut using a water-cooled diamond blade to prevent the sample's overheating. The obtained cross-sections were embedded in epoxy resin DuroFast (Struers, Copenhagen, Denmark), giving them a shape suitable for further preparation using the LaboPol-30 grinder–polisher equipped with a semi-automatic LaboForce-100 head (Struers, Copenhagen, Denmark). The preparation was carried out according to the recommendations [41]. The evaluation of the grinding and polishing process was conducted using the optical microscope NIKON MA200 (Nikon, Minato, Tokyo, Japan). The metallographic specimens were cleaned using automatic cleaning Lavamin (Struers, Copenhagen, Denmark).

The microstructure characterization was carried out on etched surfaces of metallographic specimens using the optical microscope NIKON MA200. Etching was performed for 15 s with the Kroll's reagent (Chempur, Piekary Slaskie, Poland).

The identification of phases was performed on metallographic specimens using an Empyrean X-ray diffractometer (Malvern Panalytical Ltd., Malvern, UK) with Cu-Kα (λ = 1.5406 Å) source. The study was conducted using Bragg–Brentano geometry within the 2θ angle range of 30–100°.

2.7. Microhardness Tests

The microhardness tests were conducted using the FISCHERSCOPE HM2000 microhardness tester (Helmut Fischer GmbH, Sindelfingen, Germany) with a test load of F = 0.05 N. The cross-sections of randomly selected specimens were embedded in resin, and the surfaces to be measured were prepared according to the methodology presented. To analyze local hardness values, measurements were carried out for 5 areas of the cross-section. For each area, 4 measurement points were determined, approximately 150 μm apart, as shown in Figure 3.

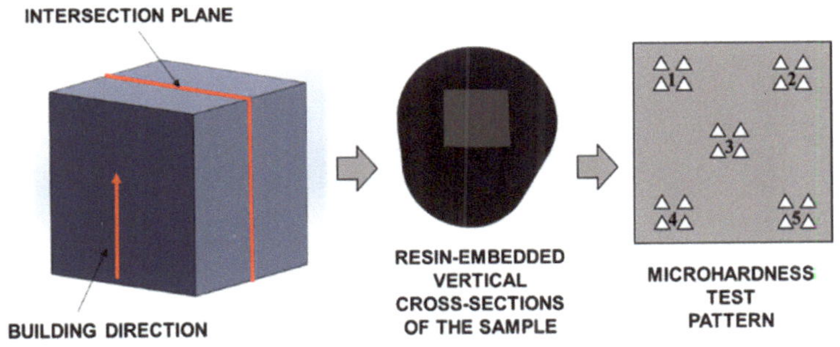

Figure 3. Scheme of preparation of metallographic sections along with microhardness test pattern (1,2,3,4 and 5-number of the area where the measurement was taken).

2.8. Uniaxial Tensile Tests

Uniaxial tensile tests were carried out on a Zwick Z400E testing machine with a macroXtens extensometer (ZwickRoell GmbH, Ulm, Germany) at ambient temperature in accordance with PN-EN ISO 6892-1:2020-05 [37]. The tests were performed for five specimens in each series. The values of tensile strength (R_m), yield strength ($R_{p0.2}$), and elastic modulus (E) were determined from stress–strain curves using the testXpert III v1.4 software (ZwickRoell GmbH, Ulm, Germany).

3. Results and Discussion

3.1. Relative Density and Structural Defects

The results of relative density measurements of samples made from Ti-6Al-4V and Ti-6Al-7Nb powders are shown in Table 3. For the Ti-6Al-4V alloy, the highest relative density value of 4.360 g/cm^3 (98.42%) was obtained, and for the Ti-6Al-7Nb alloy the value was 4.485 g/cm^3 (99.44%). In both cases, the highest relative density was observed for manufacturing strategy S8 (P = 250 W, v = 1200 mm/s, E = 70 J/mm^3).

Table 3. Relative density of Ti-6Al-4V and Ti-6Al- 7Nb fabricated by SLM.

Manufacturing Strategy Symbol	Ti-6Al-4V		Ti-6Al-7Nb	
	ρ_{AVR} [g/cm^3]	ρ_{AVR} [%]	ρ_{AVR} [g/cm^3]	ρ_{AVR} [%]
S0	4.254 ± 0.03	96.04	4.411 ± 0.03	97.80
S1	4.166 ± 0.01	94.04	4.332 ± 0.03	96.06
S2	4.282 ± 0.04	96.66	4.430 ± 0.02	98.23
S3	4.336 ± 0.02	97.88	4.462 ± 0.02	98.95
S4	4.322 ± 0.03	97.55	4.456 ± 0.02	98.79
S5	4.161 ± 0.06	93.93	4.318 ± 0.05	95.75
S6	4.225 ± 0.02	95.38	4.405 ± 0.02	97.67
S7	4.213 ± 0.03	95.10	4.375 ± 0.02	97.00
S8	4.360 ± 0.01	98.42	4.485 ± 0.01	99.44

where the values, as 100% of the material density, are 4.43 g/cm^3 for Ti-6Al-4V alloy and 4.51 g/cm^3 for Ti-6Al-7Nb alloy.

It was found that it was possible to increase the relative density of the material by increasing the volumetric energy density (Figure 4). However, for the same laser energy densities, but for different combinations of scanning speed and laser power, different material relative density values were obtained. An analysis of the effect of selected parameters shows that increasing the laser power in combination with a constant scan speed increases the relative density of the material (Figure 5A). Increasing the scan speed while keeping the laser power constant decreases the relative density of the material (Figure 5B).

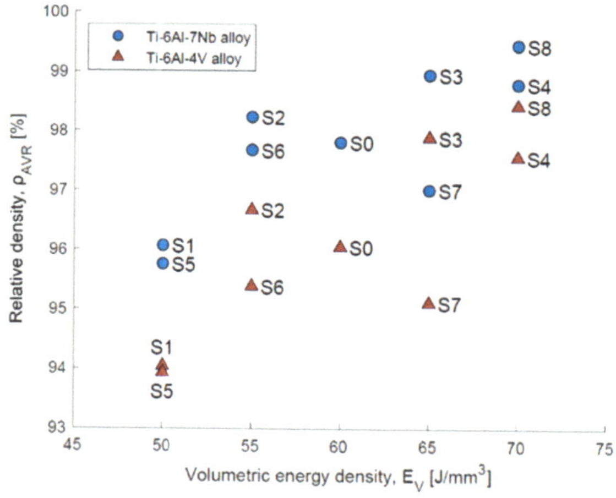

where: S0: 216 W, 1200 mm/s; **S1:** 216 W, 1430 mm/s; **S2:** 216 W, 1300 mm/s; **S3:** 216 W, 1100 mm/s; **S4:** 216 W, 1030 mm/s; **S5:** 180 W, 1200 mm/s; **S6**: 199 W, 1200 mm/s; **S7:** 233 W, 1200 mm/s; **S8:** 250 W, 1200 mm/s

Figure 4. Relative density dependent on volumetric energy density for Ti-6Al-4V and Ti-6Al-7Nb alloys.

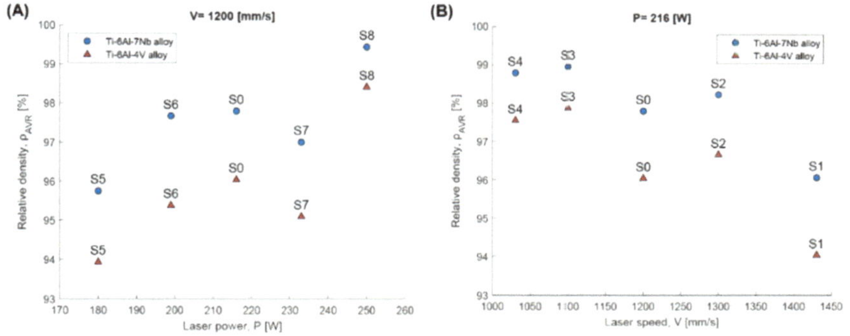

Figure 5. Relative density dependent on: (**A**) laser power; (**B**) scanning speed for Ti-6Al-4V and Ti-6Al-7Nb alloys.

The porosity of a material's internal structure is a determinant of the relative density. Figures 6 and 7 show a mosaic of metallographic images for the vertical cross-section of samples with the lowest (S5) and highest (S8) relative densities for the Ti-6Al-4V and Ti-6Al-7Nb alloys. In the samples with the lowest relative density, pores with irregular and elongated shapes were observed, which can be classified as "lack-of-fusion" defects. The pores are arranged according to the direction of layer formation, and partially fused powder grains can be observed in some of them.

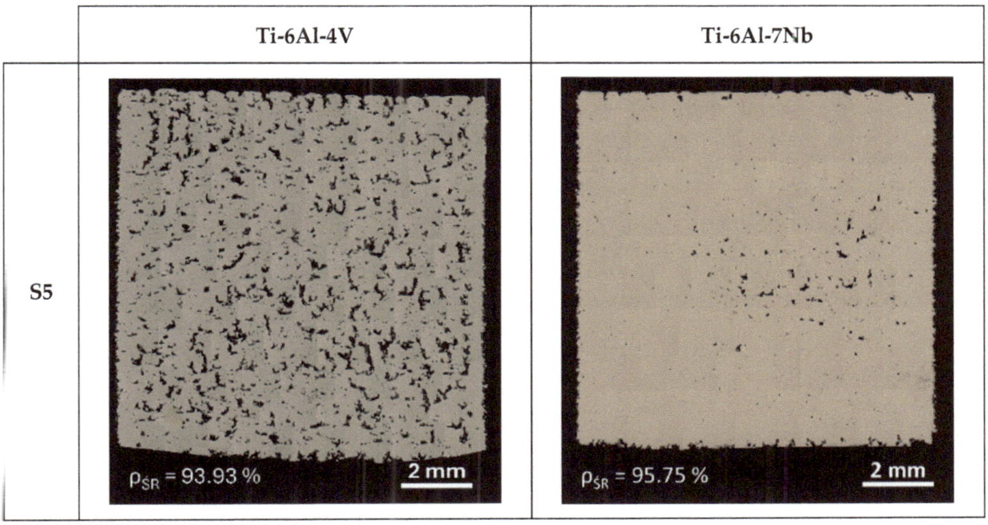

Figure 6. Mosaics of vertical cross-section of Ti-6Al-4V and Ti-6Al-7Nb samples with the lowest (S5) value of relative density.

The cause of this type of defects is, among other things, a volumetric energy density that is too low during the process. When the process is conducted with insufficient energy density, poor penetration of the molten metal pool occurs, leading to the incomplete melting of the material. Some unmelted powder grains may become trapped inside the pores. Bustillos et al. [42] demonstrated that a high scanning speed combined with insufficient laser power promotes the formation of "lack-of-fusion" defects. The studies by Liverani et al. [43] for 316 L stainless steel and by Aboulkhair et al. [44] on the AlSi10Mg alloy show that this relationship is an inherent characteristic of the SLM technology, independent of the construction material.

	Ti-6Al-4V	Ti-6Al-7Nb
S8	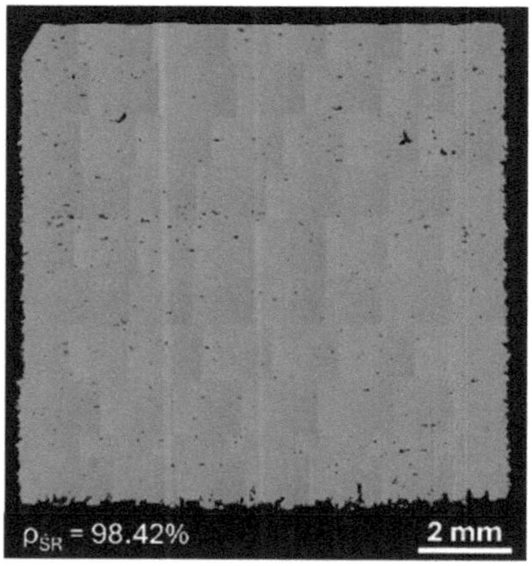	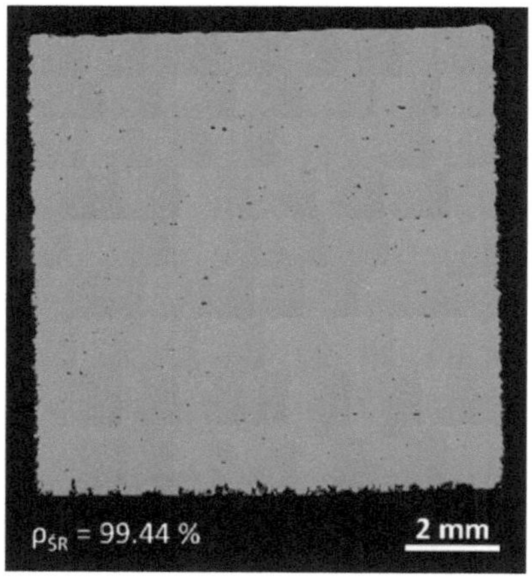

Figure 7. Mosaics of vertical cross-section of Ti-6Al-4V and Ti-6Al-7Nb samples with the highest (S8) value of relative density.

3.2. *Surface*

Figure 8 presents images of the topography of the upper surface of a sample randomly selected from each series for Ti-6Al-4V and Ti-6Al-7Nb alloys. Differences in roughness parameter values were observed for samples produced using the same manufacturing strategy, which is related to the variable orientation of the position. This is further evidence of the importance of considering the spatial orientation of individual planes in the context of their surface quality.

Table 4 shows the evaluation of surface roughness based on the values of Sa and Sz parameters (averages of three measurements). A reduction in surface roughness was obtained as the results of an increase in the value of volumetric energy density (Figure 9). The lowest average value of the Sa was observed for strategy S8. This was, respectively, 10.0 ± 3.6 μm for the Ti-6Al-4V alloy and 8.8 ± 2.0 μm for the Ti-6Al-7Nb alloy.

Table 4. Sa and Sz values for the upper surfaces of Ti-6Al-4V and Ti-6Al-7Nb samples.

Manufacturing Strategy Symbol	Ti-6Al-4V		Ti-6Al-7Nb	
	Sa [μm]	**Sz [μm]**	**Sa [μm]**	**Sz [μm]**
S0	14.3 ± 4.5	267 ± 50	16.6 ± 2.3	290 ± 25
S1	24.0 ± 3.3	350 ± 13	22.3 ± 4.4	282 ± 28
S2	13.0 ± 4.9	303 ± 61	13.0 ± 1.6	197 ± 22
S3	10.9 ± 3.2	244 ± 36	11.4 ± 2.2	209 ± 7
S4	12.7 ± 1.8	244 ±19	12.8 ± 2.2	238 ± 18
S5	26.2 ± 5.2	311 ± 52	22.5 ± 5.4	306 ± 78
S6	20.4 ± 5.2	315 ± 24	13.3 ± 1.5	253 ± 16
S7	17.4 ± 2.0	256 ± 32	20.1 ± 4.1	356 ± 41
S8	10.0 ± 3.6	184 ± 45	8.8 ± 2.0	187 ± 31

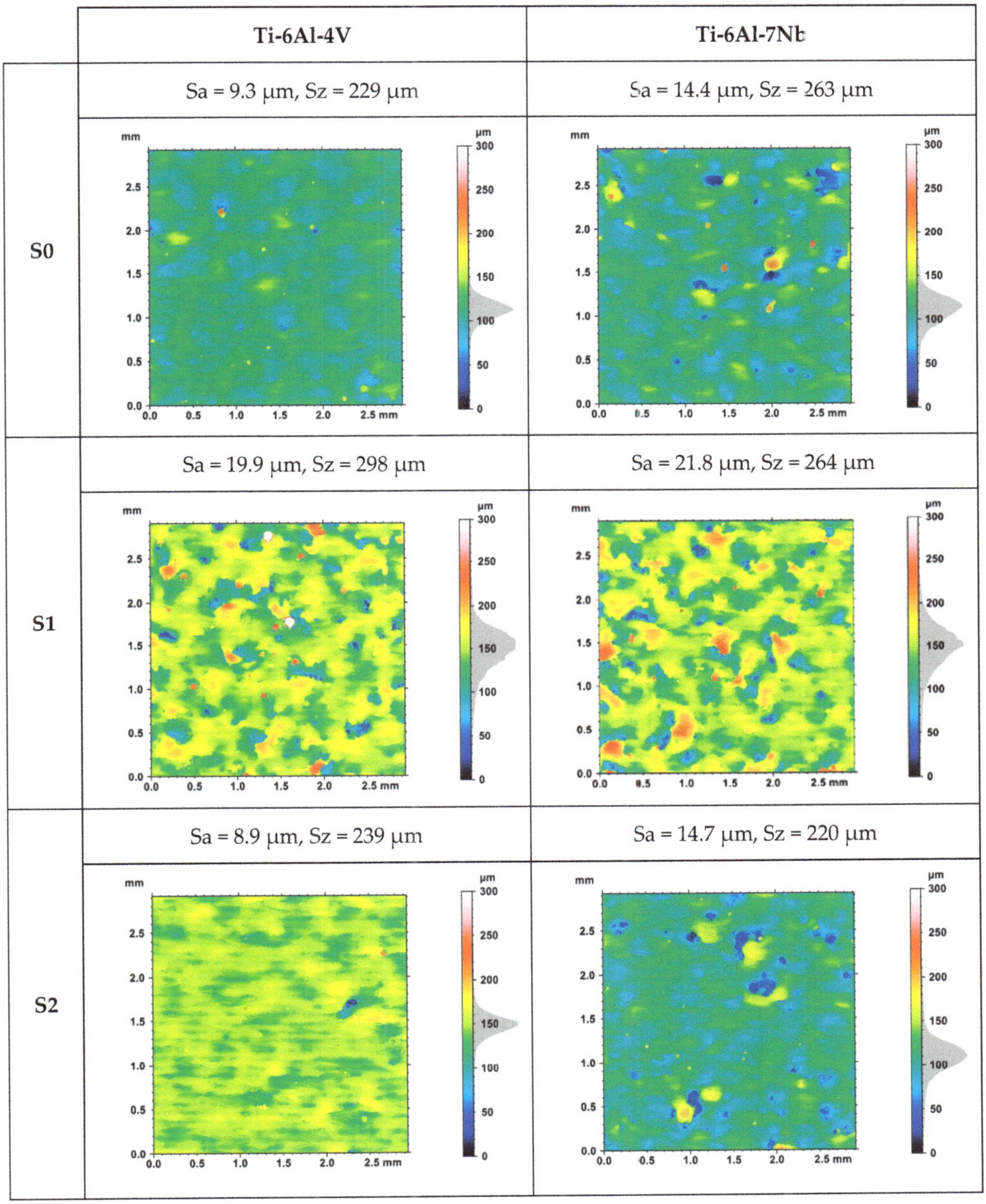

Figure 8. *Cont.*

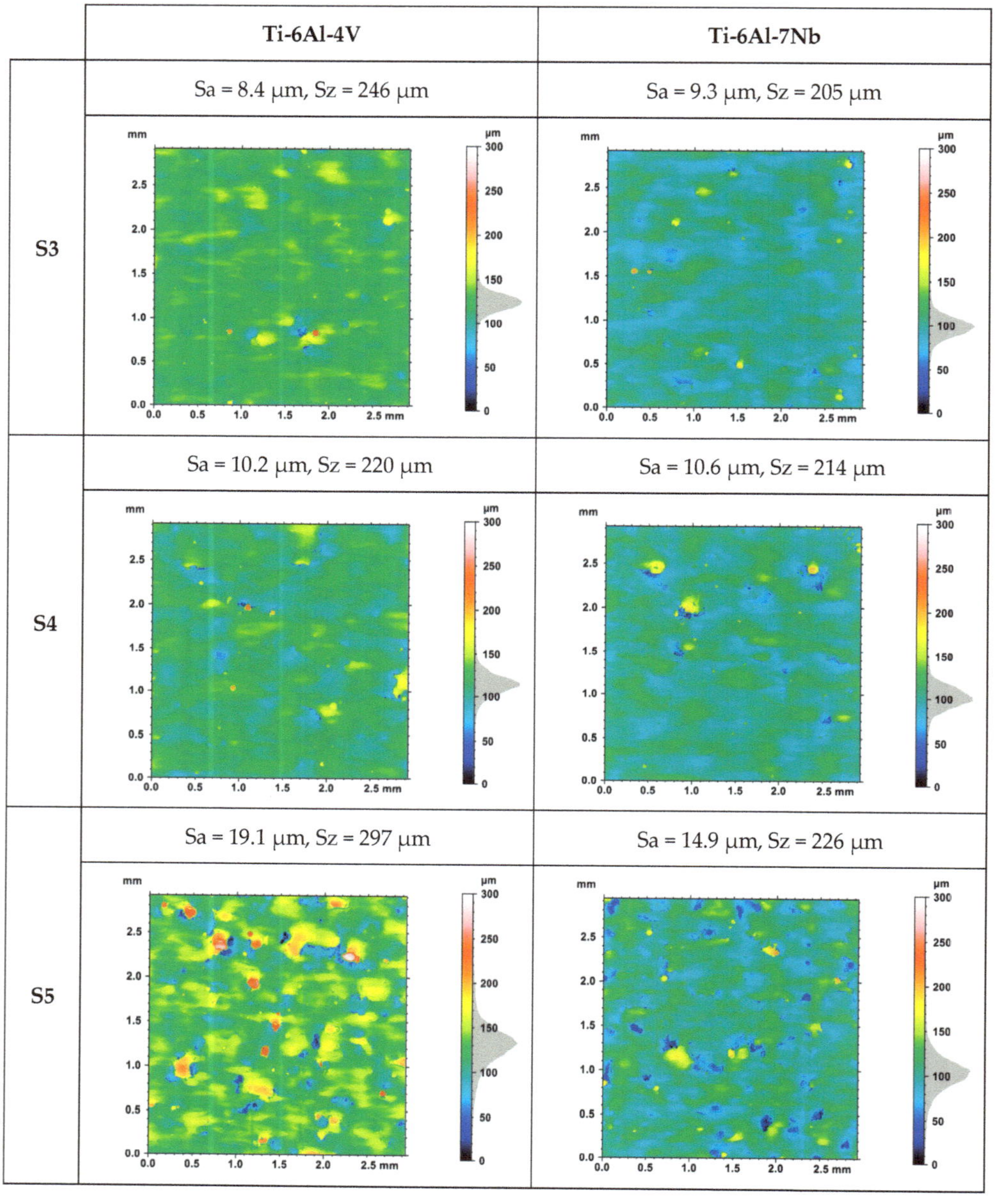

Figure 8. *Cont.*

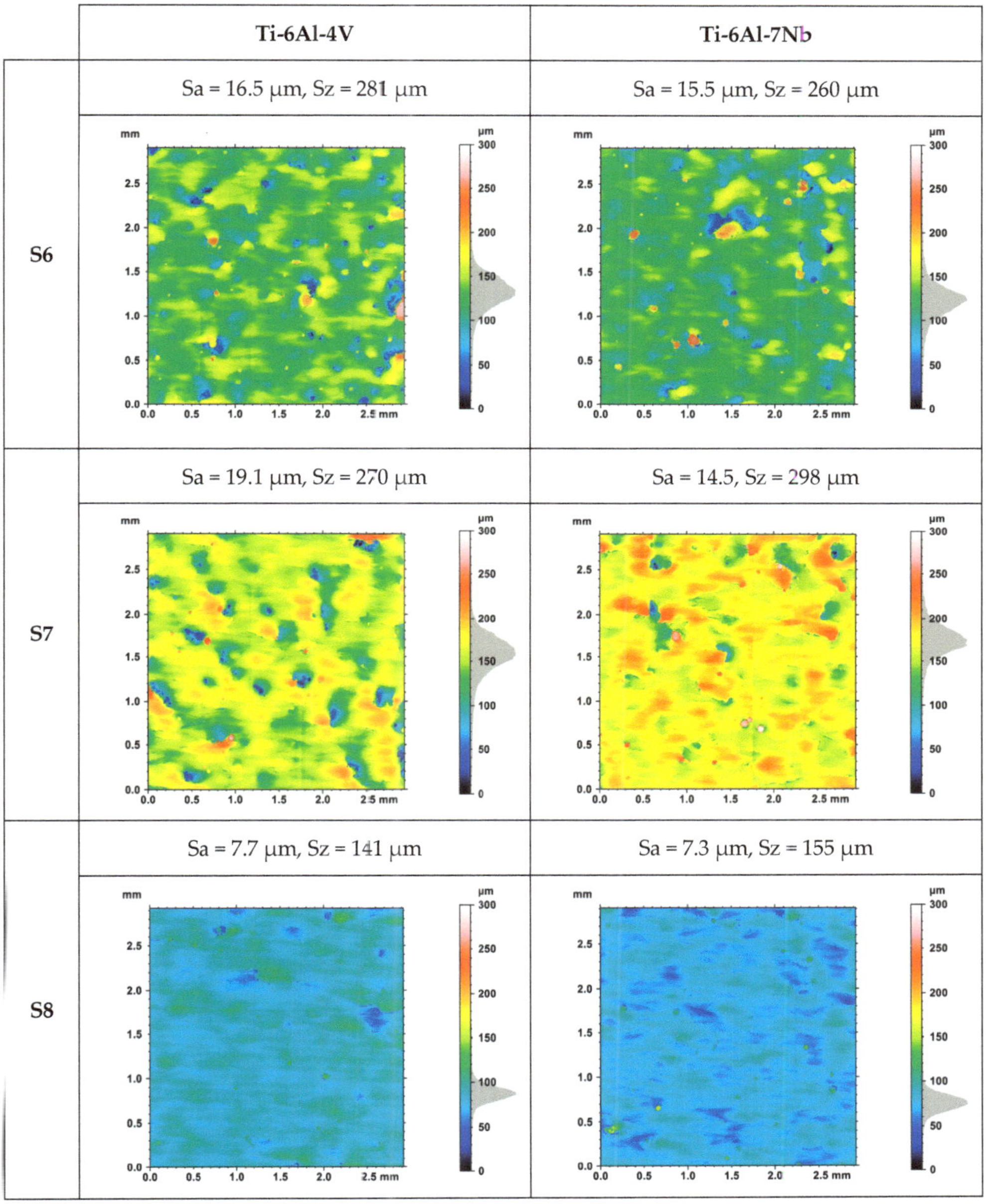

Figure 8. Topography of the top surface of randomly selected Ti-6Al-4V and Ti-6Al-7Nb samples.

The study showed a correlation of roughness with the values of laser power and scanning speed. A decrease in roughness is achieved by increasing the laser power without changing the scanning speed (Figure 10A), while an increase in roughness occurs when the laser power is held constant and the scanning speed is increased (Figure 10B).

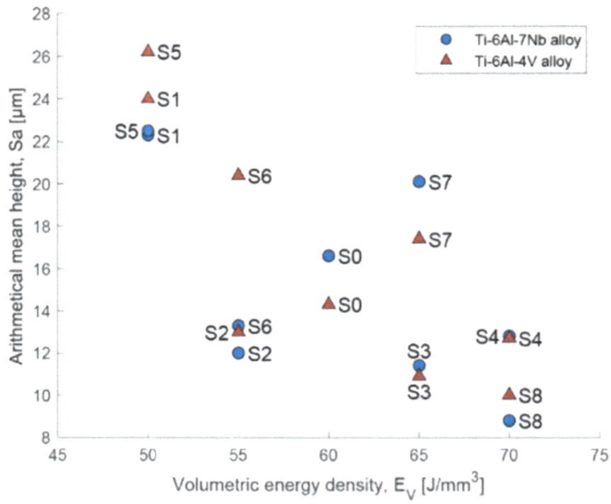

where: S0: 216 W, 1200 mm/s; **S1**: 216 W, 1430 mm/s; **S2**: 216 W, 1300 mm/s; **S3**: 216 W, 1100 mm/s; **S4**: 216 W, 1030 mm/s; **S5**: 180 W, 1200 mm/s; **S6**: 199 W, 1200 mm/s; **S7**: 233 W, 1200 mm/s; **S8**: 250 W, 1200 mm/s

Figure 9. Surface roughness defined by the parameter Sa as a function of volumetric energy density for the Ti-6Al-4V and Ti-6Al-7Nb alloys.

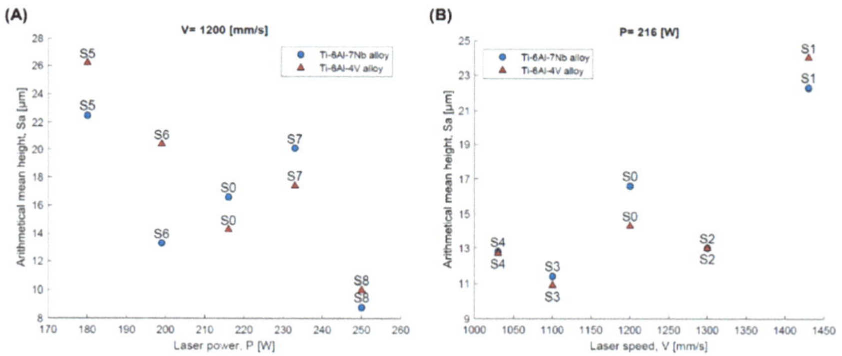

Figure 10. Surface roughness defined by the parameter Sa depending on: (**A**) laser power; (**B**) scanning speed for the Ti-6Al-4V and Ti-6Al-7Nb alloys.

Figure 11 shows the SEM images of the top surfaces of samples with the highest (S5) and lowest (S8) surface roughness for the Ti-6Al-4V and Ti-6Al-7Nb alloys.

The moving heat source (laser) creates unique marks on the top surface, known as laser welds. The welds exhibit characteristic ripples, which are the result of Marangoni convection [45]. Defects typical of SLM-manufactured surfaces were observed [46], including defects related to the so-called balling effect and the lack of complete melting of the powder. The occurrence of such defects may be related to the spatter effect and thermal diffusion. The movement of unmelted powder grains is caused by forces resulting from the surface tension gradient of the molten material. The high temperature of the melt pool causes the grains to partially melt, and when they cool, they settle at the edge of the pool [35,45]. The defects associated with the balling effect are larger and irregular in shape compared to agglomerates of unmelted or partially melted powder grains. The presence of the described defects depends on the process parameters, and their incidence increases with increasing volumetric energy density.

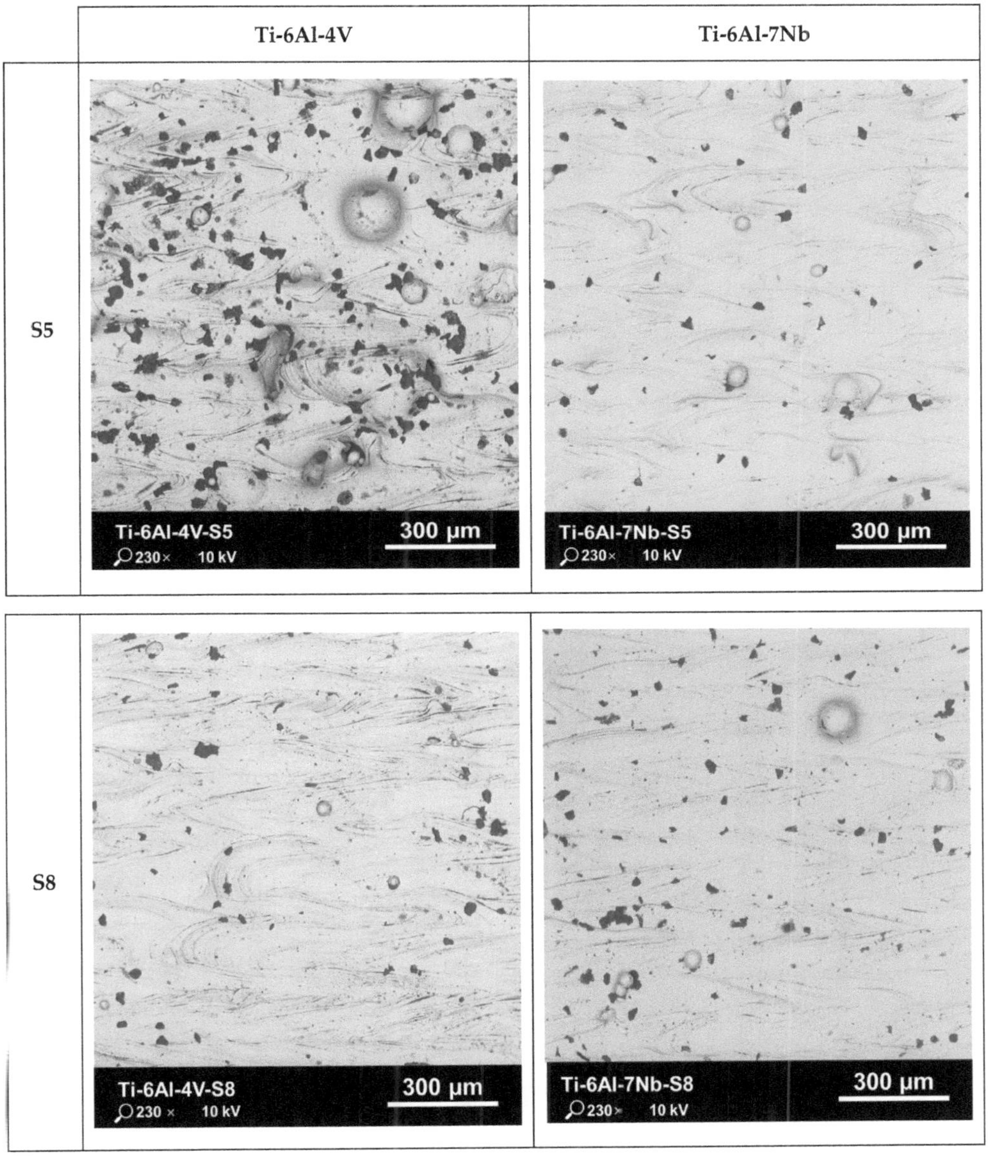

Figure 11. SEM morphology of the top surface of the Ti-6Al-4V and Ti-6Al-7Nb samples with the highest (S5) and the lowest (S8) value of surface roughness.

Increasing the surface roughness also affects the value of the determined relative density of the manufactured parts. Surface defects are not limited only to the top surface; they can also be observed on intermediate layers. Defects on the solidified surface of the intermediate layer can affect the distribution and thickness of the new powder layer. In such a situation, the delivered laser energy density may be insufficient to fully melt the material. Consequently, structural defects of the "lack-of-fusion" type may occur or be enlarged [19,27].

3.3. Microhardness

The results of the microhardness measurements of different areas of randomly selected samples made of the Ti-6Al-4V and Ti-6Al-7Nb alloy are shown in Table 5 (which occurred according to the scheme presented in Figure 3).

Table 5. Microhardness of Ti-6Al-4V and Ti-6Al-7Nb samples.

Manufacturing Strategy Symbol	$HV_{0.05}$				
	OB1	OB2	OB3	OB4	OB5
		Ti-6Al-4V			
S0	438 ± 11	426 ± 7	449 ± 4	415 ± 10	410 ± 14
S1	423 ± 8	414 ± 6	454 ± 17	418 ± 36	427 ± 8
S2	446 ± 16	424 ± 10	420 ± 5	416 ± 30	419 ± 5
S3	436 ± 4	419 ± 9	441 ± 13	436 ± 8	416 ± 4
S4	427 ± 20	418 ± 13	429 ± 9	438 ± 4	433 ± 7
S5	422 ± 10	420 ± 5	421 ± 9	411 ± 9	412 ± 13
S6	421 ± 3	424 ± 11	413 ± 8	405 ± 12	418 ± 6
S7	435 ± 6	414 ± 13	428 ± 13	426 ± 8	407 ± 26
S8	420 ± 6	411 ± 12	413 ± 18	408 ± 25	408 ± 5
		Ti-6Al-7Nb			
S0	420 ± 7	423 ± 11	428 ± 8	448 ± 6	419 ± 5
S1	402 ± 7	406 ± 10	430 ± 35	436 ± 4	395 ± 5
S2	418 ± 7	433 ± 14	406 ± 12	430 ± 5	405 ±4
S3	431 ± 7	420 ± 32	435 ± 9	433 ±8	421 ± 14
S4	442 ± 5	436 ± 11	422 ± 10	435 ± 13	398 ± 3
S5	424 ± 12	416 ± 10	428 ± 8	423 ± 8	392 ± 11
S6	416 ± 9	410 ±5	349 ± 12	424 ± 8	393 ± 6
S7	407 ± 10	435 ± 8	390 ± 10	432 ± 2	404 ± 6
S8	394 ± 9	403 ± 12	386 ± 4	380 ± 11	383 ± 1

Due to the rapid heating, melting and cooling, the various fragments of components manufactured using SLM technology are subjected to multiple heating and cooling cycles, which affects their local structure and properties [36]. This helps in differentiating the microhardness of individual areas. It was observed that areas near the edges of the section (OB1–OB4) show higher hardness than the core area (OB5). For two-phase titanium alloys, including Ti-6Al-4V and Ti-6Al-7Nb, the hardness depends on the volume ratio of the β phase to the α (martensitic) phase [47]. On this basis, it can be assumed that there is segregation and concentration of the α phase at the edges of the samples.

The data presented in Table 5 are also presented graphically (Figure 12). In the case of the Ti-6Al-4V alloy, increasing the volumetric energy density resulted in a decrease in the hardness of the core region from 407 HV to 433 HV. Similar conclusions can be found in the work of Zhao et al. [36], although the values they obtained were lower, ranging from 371 HV to 384 HV. In the case of the Ti-6Al-7Nb alloy, the change in volumetric energy density initially resulted in an increase in the hardness of the core region; a significant decrease occurred after the exceedance of 65 J/mm^3. The average core microhardness of Ti-6Al-7Nb ranged from 383 HV to 421 HV. These values are similar to those obtained by Chlebus et al. [32] (357 ± 18 HV) and Xu et al. [33] (371 ± 8 HV).

The change in scanning speed or laser power resulted in a change in the thermodynamic conditions within the melting pool. For the Ti-6Al-4V alloy, increasing the laser power did not lead to significant changes in microhardness—the measurement results for the sample produced at laser powers of 180 W (S5) and 250 W (S8) were 412 HV and 408 HV, respectively (Figure 13A). Increasing the scanning speed initially caused a decrease in the microhardness of the core area. Beyond 1200 mm/s, an increase in value was observed (Figure 13B).

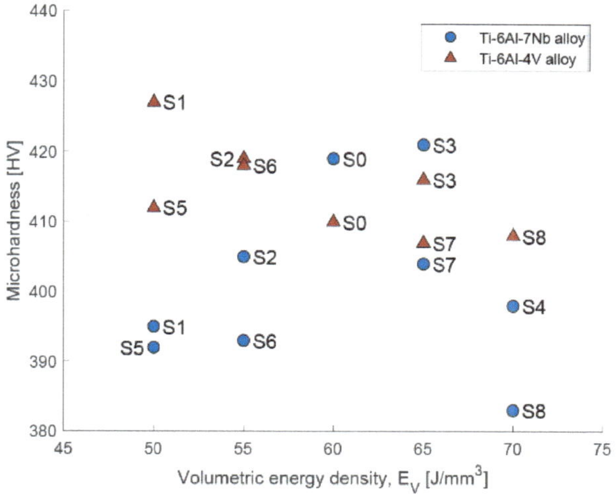

Figure 12. Microhardness of core area depending on volumetric energy density for Ti-6Al-4V and Ti-6Al-7Nb alloys.

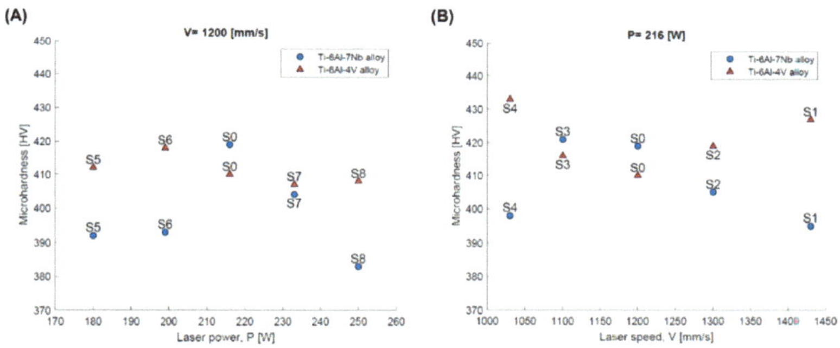

Figure 13. Microhardness of core area depending on: (**A**) laser power; (**B**) scanning speed for Ti-6Al-4V and Ti-6Al-7Nb alloys.

In the case of the Ti-6Al-7Nb alloy, an increase in laser power initially results in an increase in microhardness in the core area, with a clear maximum for the S0 strategy. Beyond 216 W, a decrease in value was observed (Figure 13A). A similar dependency was observed with increasing scanning speed, with a maximum for strategy S3. Beyond 1100 mm/s, a decrease in the microhardness of the core area was observed (Figure 13B).

Based on this, it can be assumed that the observed decrease in the microhardness of the core area was associated with an increased crystallization of the β phase in this area. The results may suggest that although both investigated alloys belong to two-phase alloys, they react differently to changes in the thermodynamic conditions prevailing in the pool of molten material caused by changes in laser power (temperature change) or scanning speed (change in heating and solidification time).

3.4. Microstructure

The as-built microstructures of the Ti-6Al-4V and Ti-6Al-7Nb alloys are shown in Figure 14. Regardless of the manufacturing strategy, the observed metallographic structure is typical for two-phase titanium alloys produced using SLM technology [48,49]. After etching, the columnar grains of the β phase, with their growth direction parallel to the build direction, were observed. The β phase crystallizes in a body-centered cubic (BCC) and is described as a solid solution of stabilizing elements, primarily V or Nb in this case [47]. The specificity of the SLM process, particularly the high solidification rates of the melting pool, promotes the crystallization of a metastable α' phase within the β phase grains. The α' phase crystallizes in a hexagonal close-packed (HCP) arrangement, and its microstructure is characterized by needle-like or plate-like features [47]. As demonstrated by the micrographs, the growth plane of the crystalline needles or plates of the α' phase is oriented at an angle of approximately 45° to the growth of the β phase grains. Given that the generation of each successive layer requires the interaction of the laser beam with a layer that has already been solidified, the occurrence of a phase transformation from α'' to β in this layer should be taken into account, as suggested by, among others, Chlebus et al. [32].

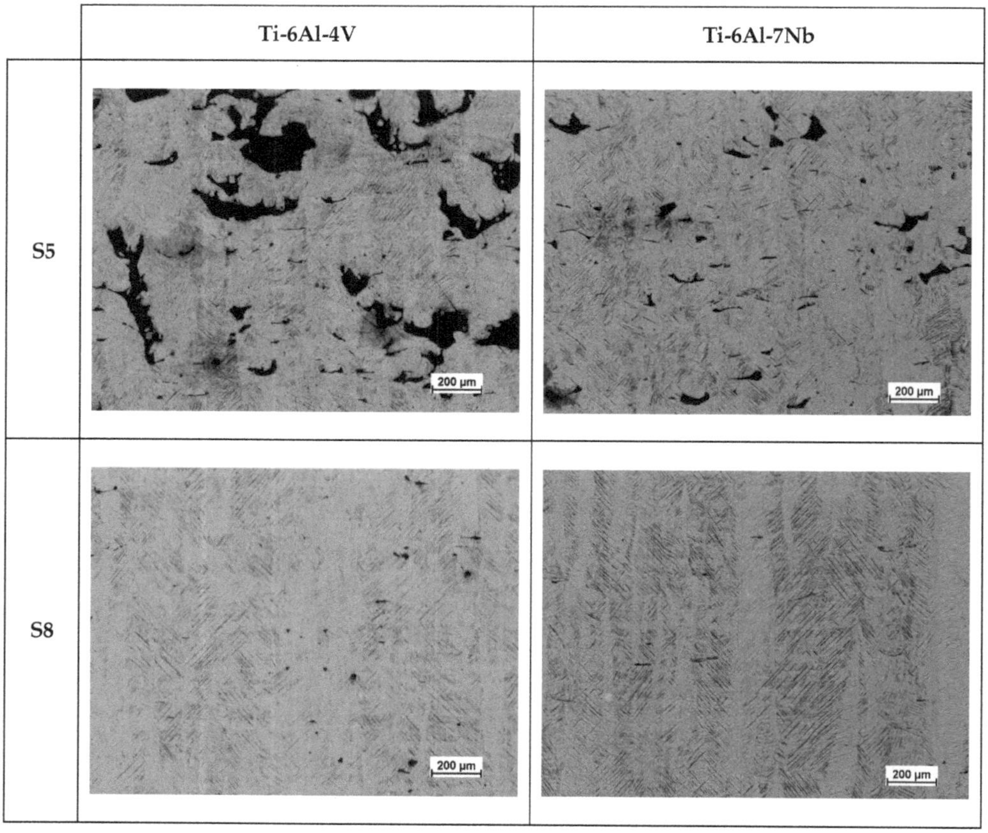

Figure 14. Microstructure of the Ti-6Al-4V and Ti-6Al-7Nb samples.

For a complete phase composition identification, X-ray diffraction analysis was conducted on samples cross-section for the Ti-6Al-4V and Ti-6Al-7Nb alloys manufactured according to strategies S5 and S8. Based on the fitting of diffraction patterns (Figure 15), it was found that for all tested alloys, the visible peaks are characteristic of the presence of titanium phase with a hexagonal close-packed (HCP) structure, due to the similarity of crystal lattice parameters [32,33,50].

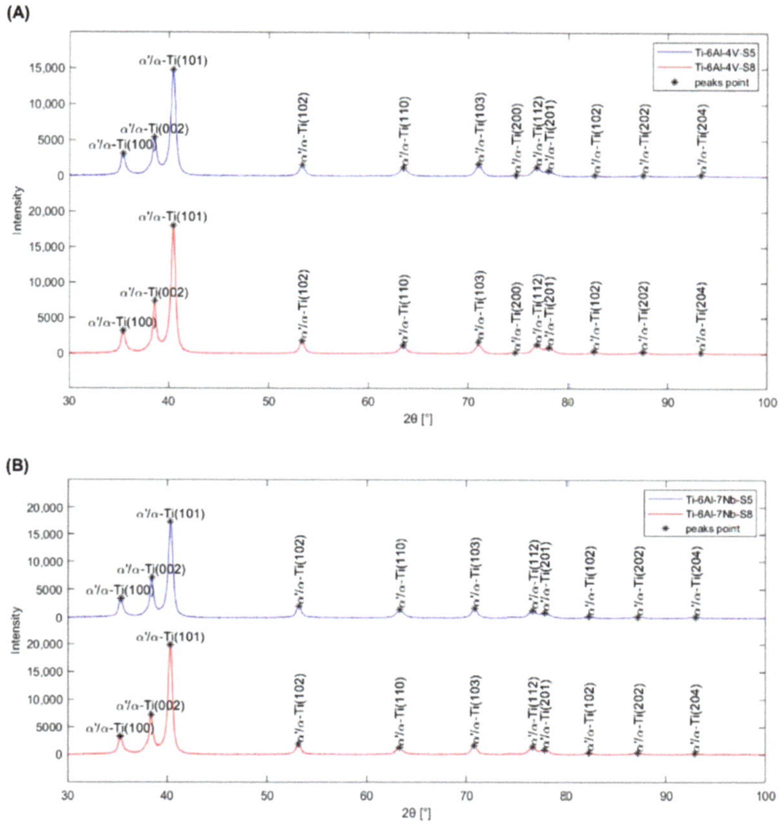

Figure 15. X-ray diffraction spectrums of (**A**) Ti-6Al-4V and (**B**) Ti-6Al-7Nb alloys manufactured according to strategies S5 and S8.

In the obtained diffraction spectrums, no additional peaks were observed. Therefore, the presence of alloying additives and thermal processing conditions did not ensure the stabilization of the β-Ti phase. The Ti-6Al-4V and Ti-6Al-7Nb alloys manufactured using SLM technology exhibit non-equilibrium, brittle microstructures with metastable α′-phase martensite. This is consistent with findings in the literature [32,33,50].

3.5. Tensile Test Results

The similarity in the stress–strain curves for samples made using the same manufacturing strategy indicates a high repeatability of the SLM process. Therefore, Figure 16 depicts the stress–strain curves of randomly selected samples from each series made from the Ti-6Al-4V and Ti-6Al-7Nb alloys. The shape of the curves is typical for materials without a distinct yield point.

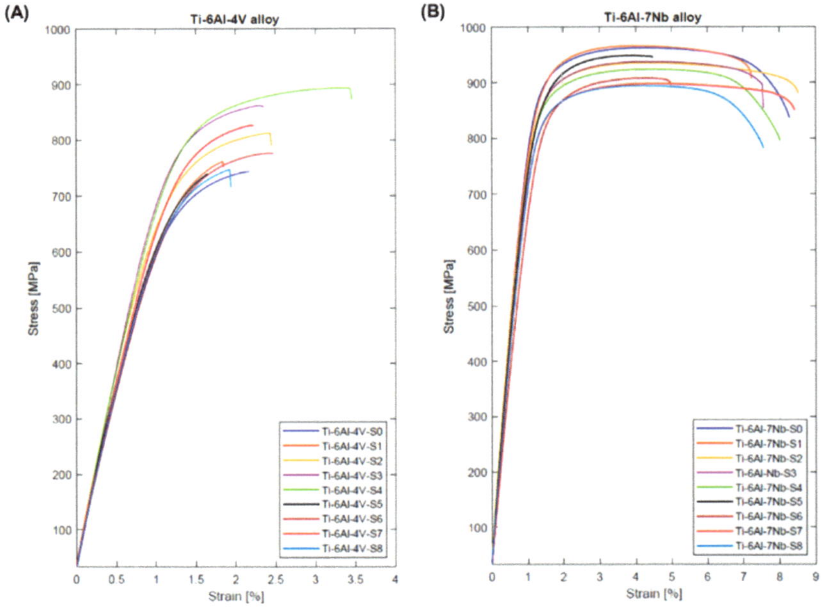

Figure 16. Stress–strain curve of (**A**) Ti-6Al-4V and (**B**) Ti-6Al-7Nb samples.

Table 6 presents a comparison of the values (an average of five measurements) of the elastic modulus (E), yield strength ($R_{p0.2}$), tensile strength (R_m), and elongation (A_{25mm}) of the Ti-6Al-4V and Ti-6Al-7Nb alloys.

Table 6. Young's modulus, yield strength, tensile strength and elongation of Ti-6Al-4V and Ti-6Al-7Nb samples.

Manufacturing Strategy Symbol	E [GPa]	$R_{p0.2}$ [MPa]	R_m [MPa]	A_{25mm} [%]
		Ti-6Al-4V		
S0	78 ± 12	565 ± 51	765 ± 22	1.2 ± 0.1
S1	67 ± 4	591 ± 22	724 ± 33	0.9 ± 0.2
S2	76 ± 4	588 ± 26	784 ± 25	1.4 ± 0.3
S3	80 ± 9	686 ± 54	860 ± 31	1.6 ± 0.5
S4	79 ± 5	726 ± 64	912 ± 17	2.2 ± 0.2
S5	73 ± 2	601 ± 20	755 ± 25	0.8 ± 0.1
S6	70 ± 4	604 ± 15	762 ± 11	1.2 ± 0.2
S7	70 ± 4	653 ± 19	801 ± 28	1.5 ± 0.3
S8	73 ± 4	649 ± 40	805 ± 38	1.3 ± 0.2
		Ti-6Al-7Nb		
S0	86 ± 5	801 ± 13	968 ± 8	7.5 ± 1.0
S1	88 ± 3	789 ± 13	977 ± 17	5.4 ± 0.6
S2	84 ± 3	779 ± 18	945 ± 20	5.7 ± 2.4
S3	83 ± 3	769 ± 13	925 ± 16	7.6 ± 0.9
S4	82 ± 4	778 ± 13	932 ± 9	7.5 ± 0.6
S5	84 ± 2	739 ± 7	932 ± 8	2.7 ± 1.1
S6	75 ± 5	720 ± 25	886 ± 24	3.3 ± 0.9
S7	81 ± 4	729 ± 25	905 ± 15	5.3 ± 1.5
S8	75 ± 3	724 ± 28	876 ± 17	6.3 ±1.5

In the case of the Ti-6Al-4V alloy, a clear increase in the values of all analyzed strength parameters was observed with an increase in volumetric energy density. A similar relationship was observed by Zhao et al. [36]. The tensile properties of the Ti-6Al-4V alloy observed in this study are lower in comparison to the range reported in previous research [36,51–53].

Conversely it was observed that, for the Ti-6Al-7Nb alloy, the tensile properties decreased with the increase in volumetric energy density. Due to the small number of literature reports, it is difficult to verify this observation. Chlebus et al. [32] reported that the tensile strength of samples from Ti-6Al-7Nb alloy with a vertical direction of layer building in relation to the working platform was 776 ± 40 MPa (the values of the other parameters are unavailable). This value is approximately 13–25% lower than that reported in this work. Unfortunately, Xu et al. p [33] did not provide clear information about the direction of layer building in the samples they analyzed, which, as is known, has a significant impact on the mechanical properties of alloys produced by SLM technology (anisotropy). However, the yield strength and tensile strength values obtained by Xu et al. were 1082 ± 13 MPa and 1160 ± 18 MPa, respectively. They are approximately 23–34% and 17–25% lower than reported in this work.

The Ti-6Al-4V alloy achieved lower values of the analyzed tensile properties parameters compared to the Ti-6Al-7Nb alloy. This is particularly evident in the case of elongation. For the Ti-6Al-7Nb alloy, the highest elongation value was 7.6 ± 0.9% and was approximately four times higher than the highest elongation for the Ti-6Al-4V alloy (2.2 ± 0.2%).

The study showed that changing the process parameters had a significant effect (laser power and scanning speed) on the analyzed mechanical properties of both alloys. In general, increasing the laser power while maintaining a constant scanning speed led to an increase in the yield strength, tensile strength, and elongation for the Ti-6Al-4V alloy. Conversely, for the Ti-6Al-7Nb alloy, a decrease in the Young's modulus, yield strength, and tensile strength values was observed, alongside an increase in the elongation value (Figure 17).

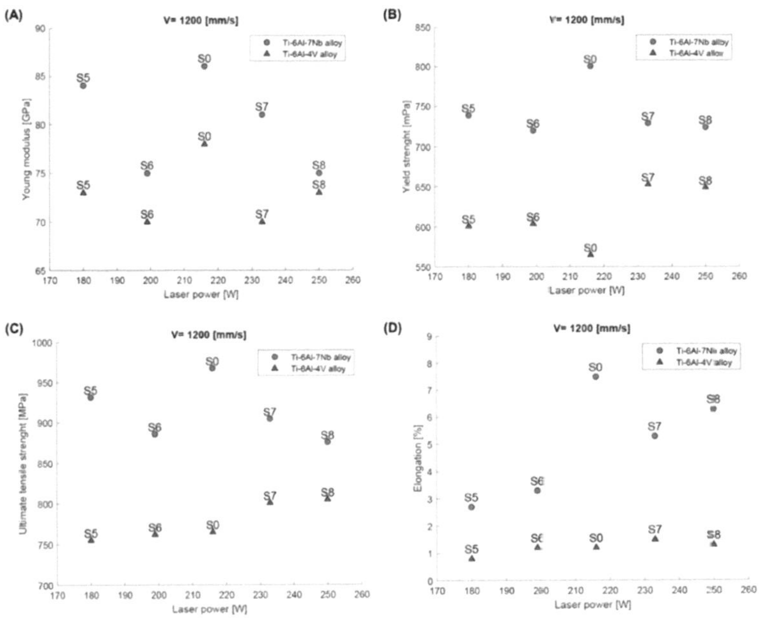

Figure 17. Mechanical properties of Ti-6Al-4V and Ti-6Al-7Nb depending on laser power: (**A**) Young's modulus; (**B**) yield strength; (**C**) tensile strength (Rm); (**D**) elongation.

4. Conclusions

The aim of the study was to determine the effect of varying the values of the process parameters (scanning speed and laser power) on the relative density, microstructure and mechanical properties of the Ti-6Al-4V and Ti-6Al-7Nb alloys. A comparison was made between the processability and properties of the two alloys produced using the same process strategies. These studies are particularly important for the Ti-6Al-Nb alloy, which is becoming increasingly popular in medicine. The currently available number of publications definitely does not exhaust the needs of the additive manufacturing industry.

For both tested materials, considering the maximization of relative density and minimization of surface roughness as selection criteria, the best strategy was S8, where the laser energy density was 70 J/mm3 with a laser power of 250 W and a scanning speed of 1200 mm/s. Regardless of the structural material used, the relative density and surface roughness can be controlled by changing the laser power or scanning speed. However, it is more advantageous to increase the laser power in order to increase the relative density while minimizing the surface roughness by adjusting the volumetric energy density.

The microstructures of the two investigated alloys were similar when they were manufactured under similar conditions. In both cases, a typical microstructure for titanium alloys manufactured by SLM technology was obtained, consisting of columnar β phase grains with a needle-like α' phase inside.

For all manufacturing strategies tested, higher relative density values were obtained for the Ti-6Al-7Nb alloy. In addition, increasing the volumetric energy density increased the tensile strength of the Ti-6Al-4V alloy, while the opposite relationship was observed for the Ti-6Al-7Nb alloy. The investigated Ti-6Al-7Nb alloy exhibited higher elongation values. The core microhardness of the Ti-6Al-7Nb alloy samples was lower than that of the Ti-6Al-4V alloy. The obtained microhardness and elongation results suggest that the Ti-6Al-7Nb alloy solidified with a higher volume of β phase.

The results show that these changes in the parameters of the fabrication process result in different material properties for the Ti-6Al-4V and Ti-6Al-7Nb alloys. The selected criteria for choosing a fabrication strategy are both practical and economical. Increased relative density in the raw state translates into better mechanical properties. Conversely, reduced roughness leads to lower time and financial costs for finishing operations. It should be noted that improvements in the microhardness and mechanical properties, including elongation, of titanium-based alloys produced by SLM technology can be achieved by appropriate heat treatment, which is a future direction of the work of the authors of this publication.

Author Contributions: Conceptualization, D.L. and B.B.; methodology, D.L. and B.B.; validation, D.L. and W.Z.; formal analysis, D.L. and W.Z.; investigation, D.L. and W.Z.; resources, B.B.; data curation, D.L.; writing—original draft preparation, D.L.; writing—review and editing, B.B.; visualization, D.L. and W.Z.; supervision, B.B. All authors have read and agreed to the published version of the manuscript.

Funding: This research received no external funding.

Institutional Review Board Statement: Not applicable.

Informed Consent Statement: Not applicable.

Data Availability Statement: The data presented in this study are available on request from the corresponding author.

Conflicts of Interest: The authors declare no conflicts of interest.

References

1. Yuan, L.; Ding, S.; Wen, C. Additive manufacturing technology for porous metal implant applications and triple minimal surface structures: A review. *Bioact. Mater.* **2019**, *4*, 56–70. [CrossRef] [PubMed]
2. Li, Y.; Yang, C.; Zhao, H.; Qu, S.; Li, X.; Li, Y. New Developments of Ti-Based Alloys for Biomedical Applications. *Materials* **201-**, *7*, 1709–1800. [CrossRef] [PubMed]

3. Niinomi, M. Mechanical biocompatibilities of titanium alloys for biomedical applications. *J. Mech. Behav. Biomed. Mater.* **2008**, *1*, 30–42. [CrossRef]
4. Eliasm, C.N.; Lims, J.H.C.; Valiev, R.; Meyers, M.A. Biomedical applications of titanium and its alloys. *JOM* **2008**, *60*, 46–49. [CrossRef]
5. Rack, H.J.; Qazi, J.I. Titanium alloys for biomedical applications. *Mater. Sci. Eng. C* **2006**, *26*, 1269–1277. [CrossRef]
6. *ASTM ISO/ASTM52900-15*; Standard Terminology for Additive Manufacturing-General Principles-Part 1: Terminology. ISO: Geneva, Switzerland, 2002.
7. Bose, S.; Ke, D.; Sahasrabudhe, H.; Bandyopadhyay, A. Additive manufacturing of biomaterials. *Prog. Mater. Sci.* **2018**, *93*, 45–111. [CrossRef]
8. Murry, L.E. Strategies for creating living, additively manufactured, open-cellular metal and alloy implants by promoting osseointegration, osteoinduction and vascularization: An overview. *J. Mater. Sci. Technol.* **2019**, *35*, 231–241. [CrossRef]
9. Chen, H.; Liu, Y.; Wang, C.; Zhang, A.; Chen, B.; Han, Q.; Wang, J. Design and properties of biomimetic irregular scaffolds for bone tissue engineering. *Comput. Biol. Med.* **2021**, *130*, 104241. [CrossRef]
10. Kanwar, S.; Vijayavenkataraman, S. Design of 3D printed scaffolds for bone tissue engineering: A review. *Bioprinting* **2021**, *24*, e00167. [CrossRef]
11. Maconachie, T.; Leary, M.; Lozanovski, B.; Zhang, X.; Qian, M.; Faruque, O.; Brandt, M. SLM lattice structures: Properties, performance, applications and challenges. *Mater. Des.* **2019**, *183*, 108137. [CrossRef]
12. Riva, L.; Ginestra, P.S.; Ceretti, E. Mechanical characterization and properties of laser-based powder bed-fused lattice structures: A review. *Int. J. Adv. Manuf. Technol.* **2021**, *113*, 649–671. [CrossRef]
13. Dong, Z.; Zhao, X. Application of TPMS structure in bone regeneration. *Eng. Regener.* **2021**, *2*, 154–162. [CrossRef]
14. Laskowska, D.; Szatkiewicz, T.; Bałasz, B.; Mitura, K. Mechanical Properties and Energy Absorption Abilities of Diamond TPMS Cylindrical Structures Fabricated by Selective Laser Melting with 316L Stainless Steel. *Materials* **2023**, *16*, 3196. [CrossRef] [PubMed]
15. Zhang, X.Y.; Fang, G.; Zhou, J. Additively Manufactured Scaffolds for Bone Tissue Engineering and the Prediction of their Mechanical Behavior: A Review. *Materials* **2017**, *10*, 50. [CrossRef]
16. Cooke, S.; Ahmadi, K.; Willerth, S.; Herring, R. Metal additive manufacturing: Technology, metallurgy and modelling. *J. Manuf. Process.* **2020**, *57*, 978–1003. [CrossRef]
17. Zheng, Z.; Jin, X.; Bai, Y.; Yang, Y.; Ni, C.; Lu, W.F.; Wang, H. Microstructure and anisotropic mechanical properties of selective laser melted Ti6Al4V alloy under different scanning strategies. *Mater. Sci. Eng. A* **2022**, *831*, 142236. [CrossRef]
18. Hitzler, L.; Hirsch, J.; Heine, B.; Merkel, M.; Hall, W.; Öchsner, A. On the Anisotropic Mechanical Properties of Selective Laser-Melted Stainless Steel. *Materials* **2017**, *10*, 1136. [CrossRef]
19. DebRoy, T.; Wei, H.L.; Zuback, J.S.; Mukherjee, T.; Elmer, J.W.; Milewski, J.O.; Beese, A.M.; Wilson-Heid, A.; De, A.; Zhang, W. Additive manufacturing of metallic components—Process, structure and properties. *Prog. Mater. Sci.* **2018**, *92*, 112–224. [CrossRef]
20. Laskowska, D.; Mitura, K.; Ziółkowska, E.; Bałasz, B. Additive manufacturing methods, materials and medical applications—The review. *J. Mech. Energy Eng.* **2021**, *5*, 15–30. [CrossRef]
21. Ngo, T.D.; Kashani, A.; Imbalzano, G.; Nguyen, K.T.Q.; Hui, D. Additive manufacturing (3D printing): A review of materials, methods, applications and challenges. *Compos. Part. B* **2018**, *143*, 172–196. [CrossRef]
22. Hebert, R.J. Viewpoint: Metallurgical aspects of powder bed metal additive manufacturing. *J. Mater. Sci.* **2016**, *53*, 1165–1175. [CrossRef]
23. King, W.E.; Barth, H.D.; Castillo, V.M.; Gallegos, G.F.; Gibbs, J.W.; Hahn, D.E.; Kamath, C.; Rubenchik, A.M. Observation of keyhole-mode laser melting in laser powder-bed fusion additive manufacturing. *J. Mater. Process. Technol.* **2014**, *214*, 2915–2925. [CrossRef]
24. Dev Singh, D.; Mahender, T.; Raji Reddy, A. Powder bed fusion process: A brief review. *Mater. Today Proc.* **2021**, *46*, 350–355. [CrossRef]
25. Rombouts, M.; Kruth, J.P.; Froyen, L.; Mercelis, P. Fundamentals of Selective Laser Melting of alloyed steel powders. *CIRP Ann.* **2006**, *55*, 187–192. [CrossRef]
26. Kruth, J.P.; Levy, G.; Klocke, F.; Childs, T.H.C. Consolidation phenomena in laser and powder-bed based layered manufacturing. *CIRP Ann.* **2007**, *56*, 730–806. [CrossRef]
27. Liu, Y.; Yang, Y.; Mai, S.; Wang, D.; Song, C. Investigation into spatter behavior during selective laser melting of AISI 316L stainless steel powder. *Mater. Des.* **2015**, *87*, 797–806. [CrossRef]
28. Neikter, M.; Huang, A.; Wu, X. Microstructural characterization of binary microstructure pattern in selective laser-melted Ti-6Al-4V. *Int. J. Adv. Manuf. Technol.* **2019**, *104*, 1381–1391. [CrossRef]
29. Widomski, C.; Solas, D.; Baudin, T.; Brisset, F.; Lebel, S.; Helbert, A.L. Heterogeneity of microstructure, texture and mechanical properties along the building direction of Ti-6Al-4V alloy developed by SLM. *IOP Conf. Ser. Mater. Sci. Eng.* **2021**, *1121*, 012031. [CrossRef]
30. Eskandari, H.; Lashgari, H.R.; Zangeneh, S.; Kong, C.; Ye, L.; Eizadjou, M.; Wang, H. Microstructural characterization and mechanical properties of SLM-printed Ti–6Al–4V alloy: Effect of build orientation. *J. Mater. Res.* **2022**, *37*, 2645–2660. [CrossRef]

31. Dareh Baghi, A.; Nafisi, S.; Ebendorff-Heidepriem, H.; Ghomashchi, R. Microstructural Development of Ti-6Al-4V Alloy via Powder Metallurgy and Laser Powder Bed Fusion. *Metals* **2022**, *12*, 1462. [CrossRef]
32. Chlebus, E.; Kuźnicka, B.; Kurzynowski, T.; Dybała, B. Microstructure and mechanical behaviour of Ti—6Al—7Nb alloy produced by selective laser melting. *Mater. Charact.* **2011**, *62*, 488–495. [CrossRef]
33. Xu, C.; Sikan, F.; Atabay, S.E.; Muñiz-Lerma, J.A.; Sanchez-Mata, O.; Wang, X.; Brochu, M. Microstructure and mechanical behavior of as-built and heat-treated Ti–6Al–7Nb produced by laser powder bed fusion. *Mater. Sci. Eng. A* **2020**, *793*, 139978. [CrossRef]
34. *PN-ISO 9276-1:1998*; Representation of results of particle size analysis—Part 1: Graphical representation. ISO: Geneva, Switzerland, 1998.
35. Dursun, G.; Ibekwe, S.; Li, G.; Mensah, P.; Joshi, G.; Jerro, D. Influence of laser processing parameters on the surface characteristics of 316L stainless steel manufactured by selective laser melting. *Mater. Today Proc.* **2020**, *26*, 387–393. [CrossRef]
36. Zhao, R.; Chen, C.; Wang, W.; Cao, T.; Shuai, S.; Xu, S.; Hu, T.; Liao, H.; Wang, J.; Ren, Z. On the role of volumetric energy density in the microstructure and mechanical properties of laser powder bed fusion Ti-6Al-4V alloy. *Addit. Manuf.* **2022**, *51*, 102605. [CrossRef]
37. *PN-EN ISO 6892-1:2020-05*; Metals, Tensile Test, Part 1: Room Temperature Test Method. ISO: Geneva, Switzerland, 2020.
38. *ISO 25178-2:2021*; Geometrical Product Specifications (GPS), Surface Texture: Areal, Part 2: Terms, Definitions and Surface Texture Parameters. ISO: Geneva, Switzerland, 2021.
39. Townsend, A.; Senin, N.; Blunt, L.; Leach, R.K.; Taylor, J.S. Surface texture metrology for metal additive manufacturing: A review. *Precis. Eng.* **2016**, *46*, 34–47. [CrossRef]
40. Triantaphyllou, A.; Giusca, C.L.; Macaulay, G.D.; Roerig, F.; Hoebel, M.; Leach, R.K.; Tomita, B.; Milne, A.K. Surface texture measurement for additive manufacturing. *Surf. Topogr. Metrol. Prop.* **2015**, *3*, 024002. [CrossRef]
41. Metallographic Preparation of Titanium, Stuers. 2024. Available online: https://publications.struers.com/brochures/english/application-notes/titanium/?_ga=2.65597550.1793077269.1699704274-2146115658.1699704274 (accessed on 15 January 2024).
42. Bustillos, J.; Kim, J.; Moridi, A. Exploiting lack of fusion defects for microstructural engineering in additive manufacturing. *Addit. Manuf.* **2021**, *48*, 102399. [CrossRef]
43. Liverani, E.; Toschi, S.; Ceschini, L.; Fortunato, A. Effect of selective laser melting (SLM) process parameters on microstructure and mechanical properties of 316L austenitic stainless steel. *J. Mater. Process. Technol.* **2017**, *249*, 255–263. [CrossRef]
44. Aboulkhair, N.T.; Everitt, N.M.; Ashcroft, I.; Tuck, C. Reducing porosity in AlSi10Mg parts processed by selective laser melting. *Addit. Manuf.* **2014**, *1–4*, 77–86. [CrossRef]
45. Cabanettes, F.; Joubert, A.; Chardon, G.; Dumas, V.; Rech, J.; Grosjean, C.; Dimkovski, Z. Topography of as built surfaces generated in metal additive manufacturing: A multi scale analysis from form to roughness. *Precis. Eng.* **2018**, *52*, 249–265. [CrossRef]
46. Thompson, A.; Senin, N.; Leach, R.K. Towards an additive surface atlas. In Proceedings of the ASPE/Euspen Conference on Dimensional Accuracy and Surface Finish from Additive Manufacturing, Raleigh, NC, USA, 27–30 June 2016; pp. 27–30.
47. Motyka, M.; Kubiak, K.; Sieniawski, J.; Ziaja, W. Phase Transformations and Characterization of α + β Titanium Alloys. In *Comprehensive Materials Processing*; Elsevier: Amsterdam, The Netherlands, 2014; pp. 7–36.
48. Thijs, L.; Verhaeghe, F.; Craeghs, T.; Van Humbeeck, J.; Kruth, J.-P. A study of the microstructural evolution during selective laser melting of Ti–6Al–4V. *Acta Mater.* **2010**, *58*, 3303–3312. [CrossRef]
49. Simonelli, M.; Tse, Y.Y.; Tuck, C. Microstructure of Ti-6Al-4V produced by selective laser melting. *J. Phys. Conf. Ser.* **2012**, *371*, 012084. [CrossRef]
50. Simonelli, M.; Tse, Y.Y.; Tuck, C. The formation of α + β microstructure in as-fabricated selective laser melting of Ti–6Al–4V. *J. Mater. Res.* **2014**, *29*, 2028–2035. [CrossRef]
51. Yan, Q.; Chen, B.; Kang, N.; Lin, X.; Lv, S.; Kondoh, K.; Li, S.; Li, J.S. Comparison study on microstructure and mechanical properties of Ti-6Al-4V alloys fabricated by powder-based selective-laser-melting and sintering methods. *Mater. Charact.* **2020**, *164*, 110358. [CrossRef]
52. Vilaro, T.; Colin, C.; Bartout, J.D. As-Fabricated and Heat-Treated Microstructures of the Ti-6Al-4V Alloy Processed by Selective Laser Melting. *Met. Mater. Trans. A* **2011**, *42*, 3190–3199. [CrossRef]
53. Yvinec, T.; Nait-Ali, A.; Mellier, D.; Bertheau, D.; Cormier, J.; Villechaise, P.; Rat, L.; Hemery, S. Tensile properties of Ti-6Al-4V as-built by laser metal deposition: The relationship between heat affected zone bands, strain localization and anisotropy in ductility. *Addit. Manuf.* **2022**, *55*, 102830. [CrossRef]

materials

MDPI

Article

Microstructure, Mechanical, and Tribological Properties of Nb-Doped TiAl Alloys Fabricated via Laser Metal Deposition

Kai Huang [1], Feng Xu [1,2], Xinyan Liu [1,2], Shiqiu Liu [3], Qingge Wang [1], Ian Baker [4], Min Song [1] and Hong Wu [1,*]

1 State Key Laboratory of Powder Metallurgy, Central South University, Changsha 410083, China
2 Farsoon Technologies Co., Ltd., Changsha 410221, China
3 Advanced Materials Additive Manufacturing Innovation Research Center, School of Engineering, Hangzhou City University, Hangzhou 310015, China
4 Thayer School of Engineering, Dartmouth College, Hanover, NH 03755, USA
* Correspondence: hwucsu@csu.edu.cn

Abstract: TiAl alloys possess excellent properties, such as low density, high specific strength, high elastic modulus, and high-temperature creep resistance, which allows their use to replace Ni-based superalloys in some high-temperature applications. In this work, the traditional TiAl alloy Ti-48Al-2Nb-2Cr (Ti4822) was alloyed with additional Nb and fabricated using laser metal deposition (LMD), and the impacts of this additional Nb on the microstructure and mechanical and tribological properties of the as-fabricated alloys were investigated. The resulting alloys mainly consisted of the γ phase, trace β_0 and α_2 phases. Nb was well distributed throughout the alloys, while Cr segregation resulted in the residual β_0 phase. Increasing the amount of Nb content increased the amount of the γ phase and reduced the amount of the β_0 phase. The alloy Ti4822-2Nb exhibited a room-temperature (RT) fracture strength under a tensile of 568 $\pm$ 7.8 MPa, which was nearly 100 MPa higher than that of the Ti4822-1Nb alloy. A further increase in Nb to an additional 4 at.% Nb had little effect on the fracture strength. Both the friction coefficient and the wear rate increased with the increasing Nb content. The wear mechanisms for all samples were abrasive wear with local plastic deformation and oxidative wear, resulting in the formation of metal oxide particles.

Keywords: TiAl alloys; Nb alloying; laser metal deposition; microstructure; mechanical properties; tribological properties

1. Introduction

The intermetallic compound TiAl, which has a density of ~4000 kg/m^3, exhibits exceptional mechanical and chemical properties at high temperatures, including high specific strength, high specific modulus, substantial creep resistance, considerable fatigue resistance, and excellent oxidation resistance [1–3]. TiAl alloys are considered the most promising candidates to replace Ni-based superalloys within the temperature range of 650~850 °C [4,5]. They have been used effectively in racing engine parts, aero-engine low-pressure turbine (LPT) blades, and in several other applications [6,7], and enable a reduction in structural weight, thereby decreasing pollutant emissions while enhancing energy efficiency [8,9]. Between 2008 and 2018, the overall manufacturing expenses for γ-TiAl alloys at the German Electrometallurgy Company dropped by almost 70%, while production surged almost thirtyfold, demonstrating their increased commercial use [10–12]. Presently, the main processing technologies for TiAl alloys include ingot metallurgy, precision casting, and powder metallurgy [13–15]. However, both their limited hot processing range and low room-temperature plasticity significantly hinder their widespread industrial use. The ingot metallurgy process is complex and encompasses melting, casting, hot isostatic pressing, and hot processing, resulting in a high rate of defective products and high costs due to significant chemical segregation [16,17]. Although LPT blades produced by precision casting have been used on a small scale, it remains challenging to manufacture components with intricate internal cavity structures [18]. On the other hand, powder metallurgy is mostly confined to laboratory research, and faces limitations regarding the size and shape of components [19].

In the past two decades, laser additive manufacturing (LAM) has emerged as the most rapidly evolving and promising technology in advanced manufacturing [20–24]. As one of the exemplary representatives of LAM technology, laser metal deposition (LMD) technology possesses distinctive features such as high energy density, exceptional processing precision, shortened processing cycles, and customized production capabilities [25,26]. Moreover, LMD technology transcends the limitations imposed by molds and size restrictions to directly manufacture parts with intricate shapes [27,28]. This enhances the design flexibility for new products while simultaneously reducing manufacturing costs [29,30].

The components created by LMD show complex microstructures and mechanical properties, which are notably different from those of conventionally cast or forged parts [31,32]. This is a result of the distinctive thermal conditions encountered throughout the LMD process, including both rapid heating (10^6–10^7 °C·s^{-1}) and cooling (>10^3 °C·s^{-1}) cycles [33,34]. TiAl alloys are brittle, and thus, susceptibility to cracking is a major challenge during their LMD processing. Recently, researchers have successfully prepared crack-free TiAl alloy blocks by adjusting the processing parameters, e.g., the laser power and auxiliary heating, to reduce cooling rates [35,36].

Most research on TiAl alloys made by LMD has concentrated on the traditional alloy Ti-48Al-2Nb-2Cr (Ti4822). For instance, Wang et al. [37] prepared Ti4822 blocks using high-power LMD that had an ultimate tensile strength (UTS) of 545 $\pm$ 9 MPa at ambient temperatures and 471 $\pm$ 37 MPa at 760 °C. The microstructure consisted of alternating columnar and equiaxed grains. Wu et al. [28] also observed this phenomenon and explained the evolution of the microstructure and the resulting change in hardness. Zhang et al. [38] studied the anisotropy of the tensile behavior of the closely related alloy Ti-47Al-2Nb-2Cr produced by LMD and obtained a UTS at a room temperature of ~650 MPa. Ti4822 is a composition system suitable for casting, but it is not obvious that it is suitable for LMD processing, which has high cooling rates and cyclic heating characteristics [39,40]. Thus, alloying elements may exhibit different effects in LMD processing compared to processing using conventional casting. Therefore, it is essential to investigate the impacts of various alloying elements on LMD-processed alloys. This will provide a basis for designing specific alloy compositions for LMD.

Nb is commonly used as a β stabilizer in TiAl alloys and extends the β-phase region to a high Al content [41,42]. This leads to an elevation in both the melt temperature (T_m) and eutectoid temperature (T_{eu}), as well as a shift of the solute temperature for the γ phase towards the Al-rich side [43]. These modifications have significant implications for refining microstructures. In addition, the addition of Nb can greatly improve the high-temperature mechanical properties via solid solution strengthening, while also enhancing oxidation resistance [44,45]. The importance of incorporating Nb has increased significantly since the development of the second generation of TiAl alloys, particularly high-Nb TiAl alloys [11]. Hence, it is crucial to explore the impacts of Nb on the fabrication of TiAl alloys via LMD.

In some practical engineering applications, including exhaust valves, turbine blades, and dispersion flaps, sliding contact is necessary for the components during their service life. This requires TiAl alloys to exhibit not only superior strength, but also good wear resistance, which necessitates an understanding of their tribological properties. Previous studies have predominantly concentrated on the friction and wear properties of TiAl alloys with low Nb content prepared through conventional processing. Wang et al. [46] investigated the effect of lamellar orientation on the friction and wear properties of directionally solidified Ti-47Al-2Nb-2Cr-2Mn and found that the highest friction coefficient was observed for a direction of 70–90°, followed by 35–50° and 0–15°, with the wear rate showing the opposite trend. Cheng et al. [47] investigated the tribological behavior of Ti-46Al-2Cr-2Nb alloys prepared by hot-press sintering at different temperatures and found that the coefficient of friction slowly decreased with increasing temperatures. Nevertheless, the friction properties of LMD-ed TiAl alloys are rarely reported.

This study employed LMD to produce TiAl alloys with different Nb contents. The microstructural evolution, tensile properties, hardness, and tribological properties of the alloys were investigated and analyzed. The aim is to offer valuable insights into the compositional design of high-performance TiAl alloys, especially those with excellent tribological properties, processed by LMD additive manufacturing technology.

2. Materials and Methods

Powders of Ti4822 and high-purity Nb (>99.6%, particle size range from 15 to 53 μm, purchased from Bright Laser Inc., Xian, China) were used in this study. The plasma rotation electrode procedure was used to create the Ti4822 powders. As shown in Figure 1a, the resulting particle size distribution, which was characterized by a Mastersizer3000 (Malvern Panalytical, Malver, UK) laser particle size analyzer, spanned 40–150 μm, with Dv(50) = 89.7 μm. X-ray diffraction (XRD) patterns, produced using Cu-K_α radiation for a diffraction angle (2θ) ranging from 20 to 90° in increments of 0.02°, were used to determine the phases present in the powder and the processed alloys. The morphology of the Ti4822 powder was characterized using an Quanta FEG250 scanning electron microscope (SEM, FEI, Hillsboro, OR, USA). Three mixed powders with different Nb contents were obtained by mixing the Ti4822 powder with the Nb powder to produce the Ti4822-xNb alloy, where x = 1 at.%, 2 at.%, and 4 at.% designated as S1, S2, and S3, respectively.

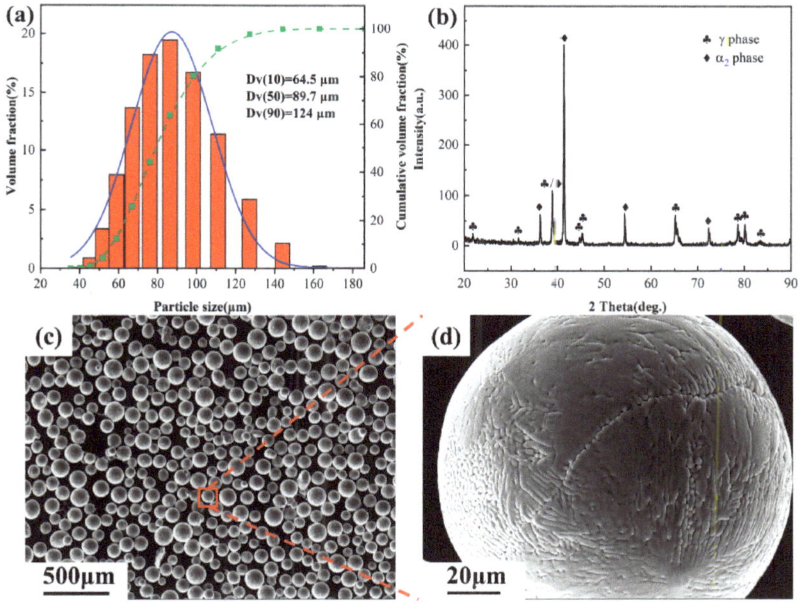

Figure 1. Ti4822 pre-alloyed powder: (**a**) particle size distribution (blue line represents the frequency fraction and green line represents cumulative fraction), (**b**) XRD pattern, (**c**) secondary electron images of the powder, and (**d**) higher magnification image from the red box in (**c**).

Ti4822-xNb samples with a size of 35 mm × 10 mm × 35 mm were produced using an RC-LDM8060 machine (Yuchen Inc., Nanjing, China) with a 2000 W fiber laser and a six-channel powder feeding system. The LMD parameters used were a laser beam power of 1600 W, a beam-scanning speed of 7 mm·s^{-1}, a layer thickness of 0.5 mm, and a beam diameter of 2.5 mm. An unalloyed Ti substrate was used for the deposition. High-purity argon gas was added to the printing chamber to keep the oxygen content below 50 ppm. A serpentine scanning strategy was adopted, as illustrated in Figure 2.

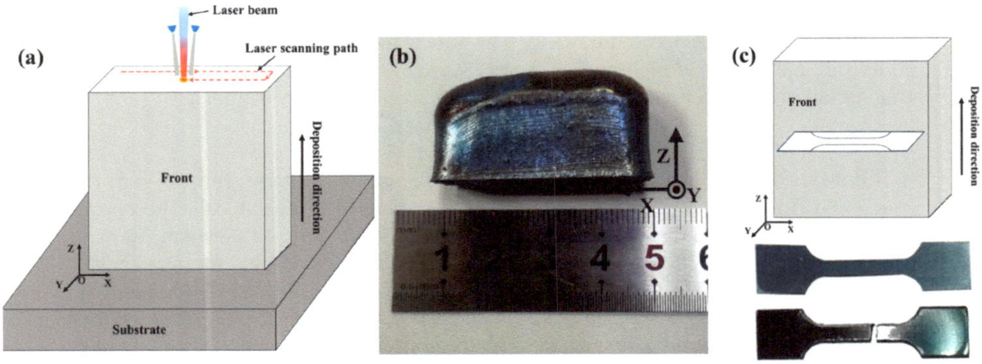

Figure 2. (**a**) Schematic of LMD processing, (**b**) photograph of as-built sample, and (**c**) sampling schematic and physical drawings of tensile samples.

The specimens were ground, polished, and then etched with Kroll's reagent (5 vol.% HF, 15 vol.% HNO_3, and 80 vol.% by volume H_2O) before examining the microstructure using a Quanta FEG250 SEM (FEI, Hillsboro, OR, USA) operated at 18 kV in the backscattered electron (BSE). The compositions of the points marked in BSE images were determined using X-ray energy dispersive spectroscopy (EDS, FEI, Hillsboro, OR, USA) at an accelerating voltage of 20 kV. The counting time for each point was 3 min. Electron backscatter diffraction (EBSD) examinations were performed at an operating voltage of 20 kV and a step size of 0.5 μm. Hardness tests were conducted using a Buehler5104 microhardness tester (Future Tech, Kanazawa, Japan) with a maximum indentation load of 0.3 kg and a duration of 20 s. Dog bone-shaped tensile specimens with gauges measuring 9.5 mm × 2 mm × 1.6 mm were cut parallel to the XY plane. Tensile tests were conducted at room temperature using a universal tensile machine (AGS-X, Shimadzu, Kyoto, Japan) with an initial strain rate of $1 \times 10^{-4}\ s^{-1}$. Three samples were measured for each alloy, and the average was calculated. An HT-1000 friction and wear testing apparatus (Zhongke Kaihua Co., Ltd, Lanzhou, China) was used to conduct dry-sliding wear tests at room temperature. The samples were rotated by sliding against a 4 mm diameter Si_3N_4 ceramic ball (purchased from Sikeer Industry & Trade Co., Ltd, Kaifeng, China) for 30 min at a rotation speed of 560 r/min with a contact load of 1 kg. The wear marks on the samples after testing were observed and imaged using a VHX-5000 ultra depth of field 3D microscope system (Keyence Co,. Ltd., Shanghai, China).

3. Results and Discussion

This section will present and analyze the results of XRD, SEM, EBSD, tensile testing, and friction and wear testing for S1, S2, and S3 samples.

3.1. Microstructure

XRD patterns revealed that the strongest diffraction peak from the Ti4822 powder corresponded to the α_2 phase (201), see Figure 1b, indicating that the powder matrix was the α_2 phase rather than the γ phase. This is because of the high cooling rate during powder processing. Figure 1c,d shows the morphology of the Ti4822 powder, which had good sphericity, dispersion, and surface finish, thereby meeting the requirements for LMD manufacturing. XRD patterns of S1, S2, and S3 are shown in Figure 3. The predominant diffraction peaks in all three samples corresponded to the γ phase (111). Most other diffraction peaks corresponded to the γ phase, while a few corresponded to the α_2 phase, indicating that the samples were mostly composed of the γ phase with a small amount of

α_2 phase. Many X-ray peaks were bifurcated, because of the tetragonal crystal structure of the γ phases [38]. Note that S1 also had a diffraction peak corresponding to the β_0 phase, which was not present in the other two samples. In addition, as the Nb content increased, the intensity of the diffraction peaks corresponding to the α_2 phase decreased, while those corresponding to the γ phase increased, indicating reduced α_2 content and increased γ content. Figure 3b shows an enlarged section of the XRD patterns around the γ (111) peak. It is evident that the diffraction peak corresponding to the (111) crystal plane of the γ phase shifted slightly to the left with increasing Nb, i.e., the 2θ value decreased. Thus, from Braggs law [48],

$$2d\sin\theta = \lambda \tag{1}$$

where d is the crystal plane spacing, θ is the diffraction angle, and λ is the wavelength of the incident X-ray. A decrease in the value of θ implies an increase in the value of d, which indicates an increase in the lattice parameter of the γ phase. The lattice parameters increase when Nb replaces Ti in TiAl alloys because Nb has a larger atomic radius than both Ti and Al. Therefore, it is evident that the amount of Nb content in the γ phase increased.

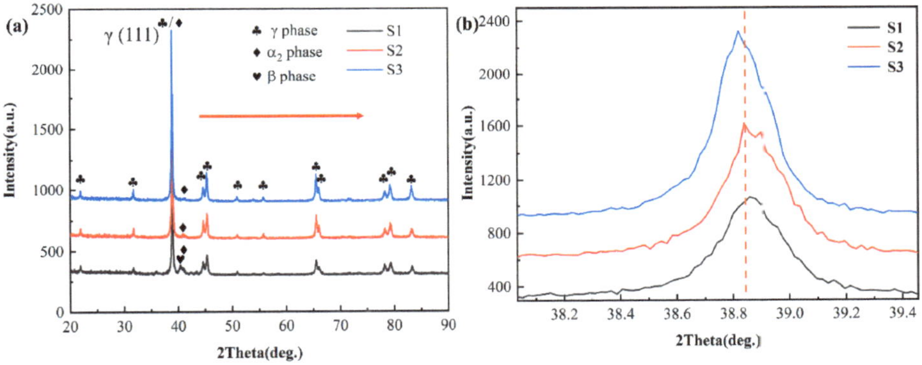

Figure 3. (**a**) XRD patterns of Ti4822-xNb alloys S1, S2 and S3, and (**b**) enlarged XRD patterns around the γ (111) peak.

Figure 4 shows backscattered electron (BSE) images of the three LMDed Ti4822-xNb alloys. No unmelted Nb particles or significant Nb segregation were observed in the samples. This suggests that the Nb particles introduced as element powders were well melted and integrated during the deposition process. Due to its low atomic number, the Al will show dark contrast in BSE images. There were obvious black network bands in the samples (as yellow arrows designated), which were regions of segregation caused by the accumulation of Al, also known as S-segregation [49]. The β stabilazing elements have a higher atomic number, so the β_0 phase appears bright white. The microstructures of all three samples exhibited S-segregation, with β_0 phases dispersed throughout the matrix. As the Nb content increased, the S-segregation became more severe while the amount of the β_0 phase decreased. To further clarify the phases present and their distribution, the chemical compositions of different regions of the samples were analyzed, see Table 1. The β_0 phase in the three samples can be divided into two types. As shown in Figure 4b, the first type was larger and had a lower Al content (location A). The second type was finer and precipitated in the middle of the lamellae (locations C, E, and I in Figure 4b,d,f). Overall, Nb did not exhibit significant segregation, and the β_0 phase was enriched in Cr. The degree of Cr enrichment in the β_0 phase of S1 was notably greater than in S2 and S3.

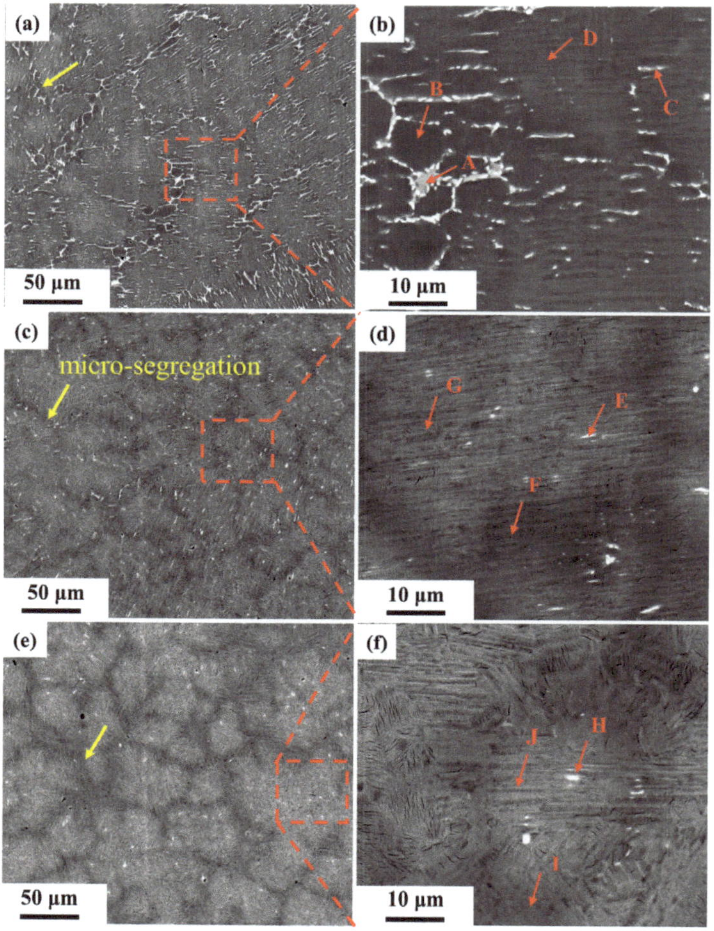

Figure 4. BSE images of Ti4822-xNb alloys (the red letters indicate features discussed in the text): (**a**) S1, (**b**) higher magnification image of the region in the red box in (**a**), (**c**) S2, (**d**) higher magnification image of the region in the red box in (**c**), (**e**) S3, (**f**) higher magnification image of the region in the red box in (**e**).

Table 1. EDS results of the marked areas in Figure 4.

Label	Chemical Composition (at.%)				Identified Phase
	Ti	Al	Nb	Cr	
A	54.1	35.6	2.0	8.3	β_0
B	44.1	52.5	2.2	1.2	γ
C	48.4	46.2	2.5	3.0	β_0
D	47.7	47.8	2.6	1.9	γ/α_2
E	45.9	46.5	4.6	3.0	β_0
F	44.7	49.6	3.3	2.4	γ/α_2
G	45.8	47.8	4.3	2.0	γ/α_2
H	44.8	47.4	4.1	3.7	β_0
I	44.8	49.4	3.7	2.1	γ/α_2
J	45.4	48.0	4.7	1.9	γ/α_2

EBSD results from the three alloys are shown in Figure 5. From the inverse pole figures (Figure 5a–c), S1 exhibited a large number of massive grains, while S2 and S3 samples showed lamellar colonies. Note that the lamellar colony size of S3 was smaller than that of S2 because the greater Nb content refines the microstructure. There was no significant texture in any of the samples. Table 2 shows the volume fractions of the phase in the three alloys obtained from EBSD results. S1, S2, and S3 contained 92.8%, 97.7%, and 98.6% of the γ phase, respectively, the volume fractions of the β_0 phase were 5.7%, 1.9%, and 1%, respectively, while the volume fractions of the α_2 phase were 1.5%, 0.4%, and 0.4%, respectively. That the volume fraction of the β_0 phase decreased with the increasing in Nb content is consistent with the BSE images. As shown in Figure 5d,e, the β_0 phases in S1 were mainly located at the grain boundaries, whereas in S2 and S3, they were located in the middle of the lamellae. Furthermore, fine granular grains crystal clusters were present along the edges of the lamellar colonies, as illustrated in Figure 5e,f. Zhang et al. [38] and Wu et al. [28] also observed this phenomenon in Ti4822 alloys formed by LMD. This is because the laser repeatedly swept over the deposited material, causing the first deposited part to undergo solid phase transitions due to thermal cycling during LMD processing [28,37]. Figure 5g–i shows that the kernel average misorientation (KAM) of the samples increased with increasing Nb, which reflects the accumulation of local faults and stresses. A large amount of Nb in a solution causes lattice distortion in TiAl alloys, leading to stress concentrations. Additionally, Nb reduces the stacking fault energy of the γ phase and increases the tendency to form mechanical twins [50].

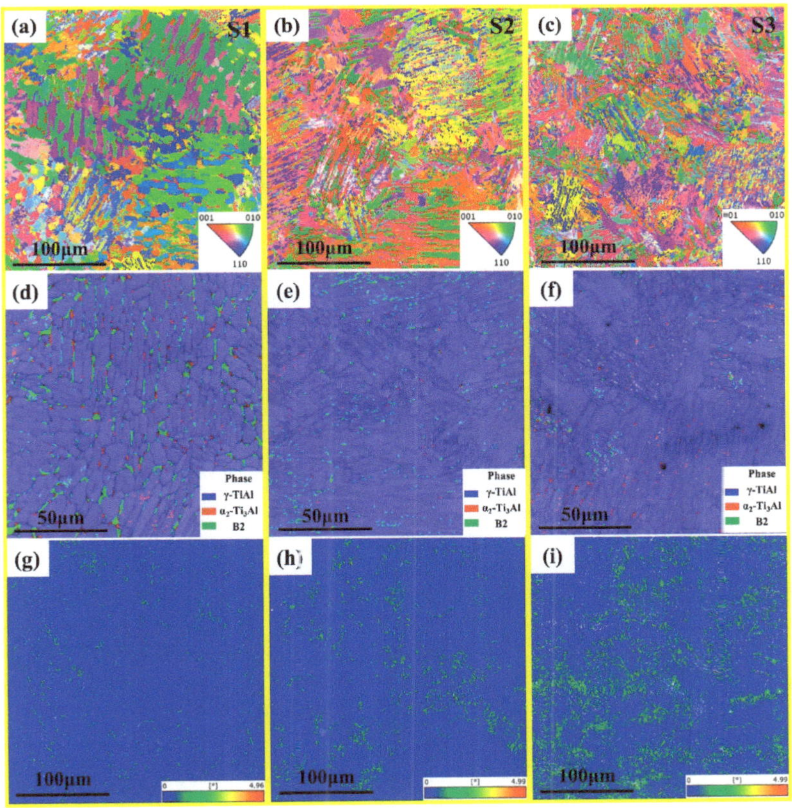

Figure 5. EBSD results including inverse pole figures (IPFs), phase maps, and kernel average misorientation (KAM) maps of Ti4822-xNb alloys: (**a**,**d**,**g**) S1, (**b**,**e**,**h**) S2, and (**c**,**f**,**i**) S3.

Table 2. Volume fractions of phases in S1, S2, and S3 from EBSD results.

Alloys	Phase Content (%)		
	γ-TiAl	β_0/B2	α_2-Ti_3Al
S1	92.8	5.7	1.5
S2	97.7	1.9	0.4
S3	98.6	1	0.4

3.2. Microstructural Evolution Analysis

Based on the solidification pathways, TiAl alloys can be divided into two categories: peritectic-solidified TiAl alloys (PSG) and β-solidified TiAl alloys (BSG). Traditional PSG alloys typically have an Al content of over 45 at.% and experience a peritectic reaction (L + $\beta \rightarrow \alpha$) during the cooling process [51]. Ti4822 is a typical PSG alloy. The above results indicate that the microstructure of S1 significantly differed from those of S2 and S3, probably due to its different solidification path. The chemical composition of S1 was closely similar to that of the base Ti4822 alloy, leading to the peritectic solidification. Figure 6a depicts the corresponding microstructural evolution of peritectic solidification of the TiAl alloy. During the LMD process, the powders were irradiated by a high-energy laser beam, leading to a local temperature spike instantaneously above the melting point. A micro-scale molten pool was created in a short period. The primary β phase nucleated and grew from the molten liquid phase. Subsequently, the α phase formed on the surface of the primary β phase because of the lower energy of nucleation and consumed both the liquid and β phases during growth. Simultaneously, the diffusion of Al from the β phase into the α phase occurred, while β-stabilizing elements diffused into the β phase. This trend led to an enrichment of β-stabilizing elements and an absence of Al at the junction of β and α grains. The cooling rate was so rapid that the alloy underwent non-equilibrium solidification, resulting in β_0 (location A in Figure 4b) + γ (location B in Figure 4b) formation. The primary β phase is a heterogeneous nucleation site which enhances the nucleation rate and generates numerous refined α grains [52]. The β_0 phase at the grain boundary is conducive to limiting the growth of α grains during subsequent thermal cycles. If α grains have an opportunity to coarsen afterward, the segregation of the β-stabilizing elements is transferred to the inner lamellae of the grains, see location C in Figure 4b.

Both the alloying elements and the cooling rate during solidification will affect the solidification path [11]. The larger Nb addition in the S2 and S3 samples further expanded the β field area to encompass higher levels of Al content, resulting in the occurrence of β-solidification, as illustrated in Figure 6b. The presence of significant S-segregation (Figure 4c,e) in both microstructures provides compelling evidence for this. The phase transition passes through a single-phase β field, which leads to the solidification of the β phase first and pushes the Al to the grain boundaries, producing S-segregation. At the phase transformation point of $\beta \rightarrow \alpha + \beta$, the β-stabilizing elements are expelled from both the boundary and the α/β interface. When the transition from β to α is incomplete, the segregation site tends to form a β_0 phase, which is dependent on the Al content present [49]. As previously described, the laser rapidly melts the powders to form a miniature molten pool. However, due to the short time, the Nb powders tend to form atomic clusters after melting [53]. On the one hand, Nb can increase the liquidus temperature of TiAl alloys, thereby promoting nucleation in the Nb-rich region. On the other hand, the Nb atomic clusters can act as heterogeneous sites to promote nucleation, thus refining the primary β grains. Consequently, in comparison to S2, S3 had finer lamellar colonies, although this also leads to a more severe S-segregation.

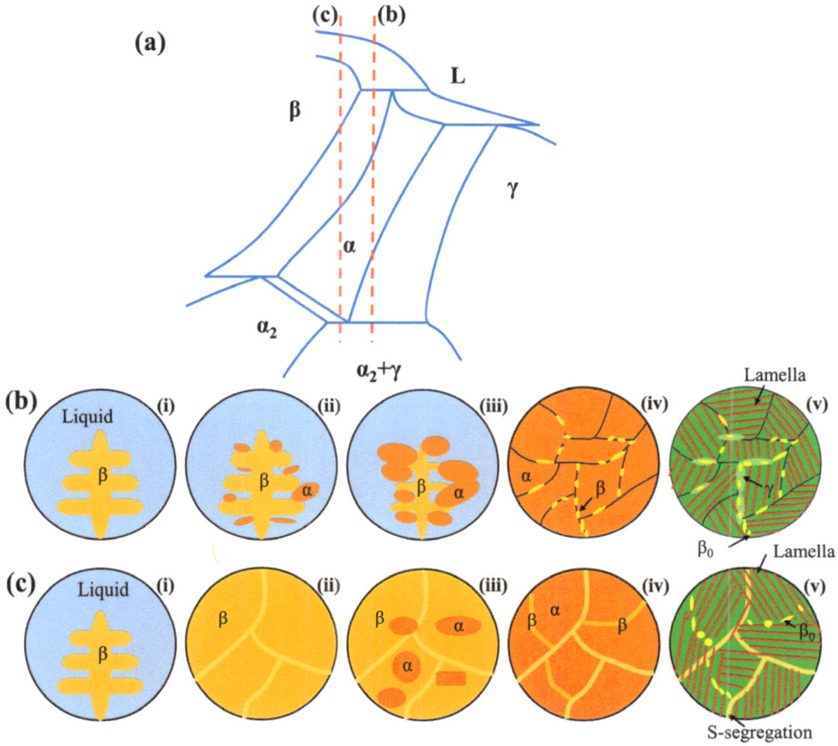

Figure 6. Schematic of the phase transformation in Ti4822-xNb alloys: (**a**) generalized Ti-Al binary phase diagram, (**b**) solidification pathway of S1, and (**c**) solidification pathways S2 and S3. (**b**,**c**) represent phase transformation during one cycle of LMD.

The above results show that the Nb element is less prone to segregation during LMD processing, which can be ascribed to the high cooling and the low diffusion coefficient. The segregation of Cr is the reason for the residual $β_0$ phase at room temperature. Adding Nb can alter the solidification path of the LMD-ed TiAl alloys, which can consequently affect their microstructures. Compared to the peritectic solidification path (S1), β-solidification (S2 and S3) is advantageous for reducing Cr segregation. This is because compared to the solid phase, the solute diffuses faster in the liquid phase [54]. Additionally, a significant amount of Nb in a solid solution can decrease the diffusion efficiency rate of Cr atoms.

3.3. Mechanical Properties

The microhardness results for S1, S2, and S3 are shown in Figure 7a. Five points were measured for each. The microhardness first increased and then decreased with increasing Nb content, with the S2 sample having the highest microhardness of 359 ± 6 HV0.3. Typical RT tensile engineering stress–strain curves are presented in Figure 7b, and the tensile results (which are the average of three tests) are presented in Figure 7c. Compared to S1, whose fracture strength was 470 ± 30 MPa, S2 exhibited an increase of nearly 100 MPa, reaching 568 ± 8 MPa. The fracture strengths of S2 and S3 samples were similar, indicating that the additional Nb did not effectively improve the strength. All three samples showed brittle fracture with elongations less than 0.5%.

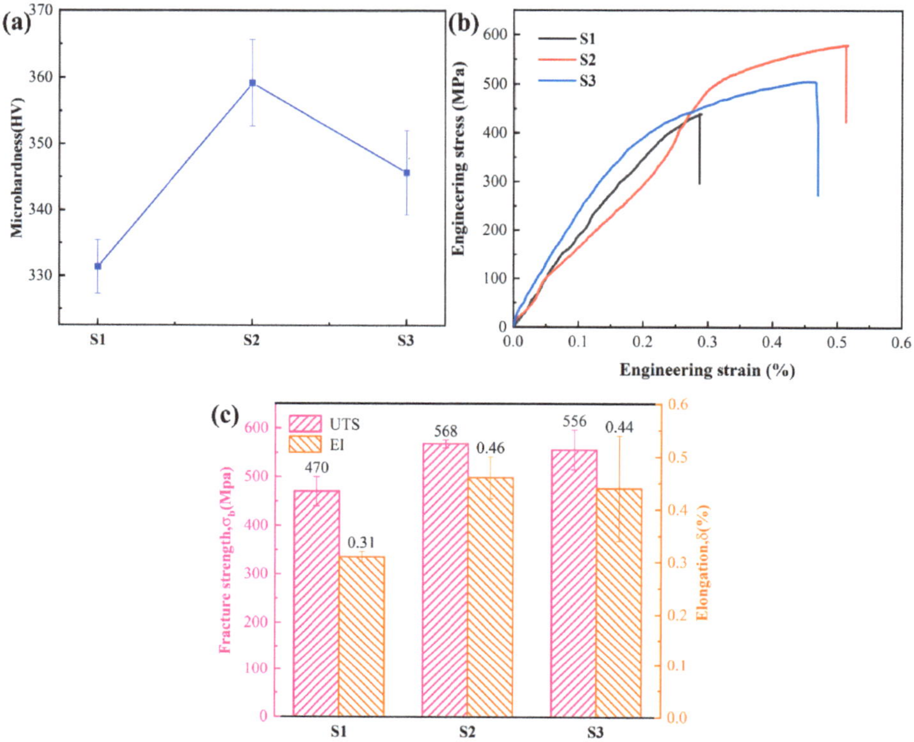

Figure 7. Mechanical properties of Ti4822-xNb alloys: (**a**) microhardness, (**b**) RT tensile engineering stress-strain curve, and (**c**) RT tensile strength and elongation.

Figure 8 shows the fracture surfaces of the three alloys after tensile testing. All the alloys exhibited brittle fracture behaviors, as evidenced by the cleavage facets. In Figure 8b,d,f, point-radiant trans-granular expansion (the red arrows) across multiple lamella thicknesses can be found. According to the fracture theory, the point of expansion is the source of cracks.

The mechanical properties of the TiAl alloys depend on two factors. First, Nb is present in a solid solution in the TiAl matrix. This not only alters the phase transformation interval but also strengthens the alloy. The addition of Nb increases the c/a ratio of the γ phase, makes it more anisotropic, and increases the strength of the γ phase [44]. Therefore, the hardness and strength of S2 and S3 were greater than those of S1. However, increasing the Nb content resulted in slightly lower hardness values for S3. This is because lattice distortion increases the brittleness of the alloys. Second, the mechanical properties are significantly affected by the phases present [55]. The γ phase has a face-centered tetragonal structure with many independent slip systems. The α_2 phase has a close-packed hexagonal structure, and its slip system is limited. The β_0 phase, which has an ordered body-centered cubic structure, is a hard and brittle phase at room temperature and is detrimental to the ductility. Therefore, the improved properties of the S2 sample also benefited from the increment of the γ phase and the reduction in the β_0 phase.

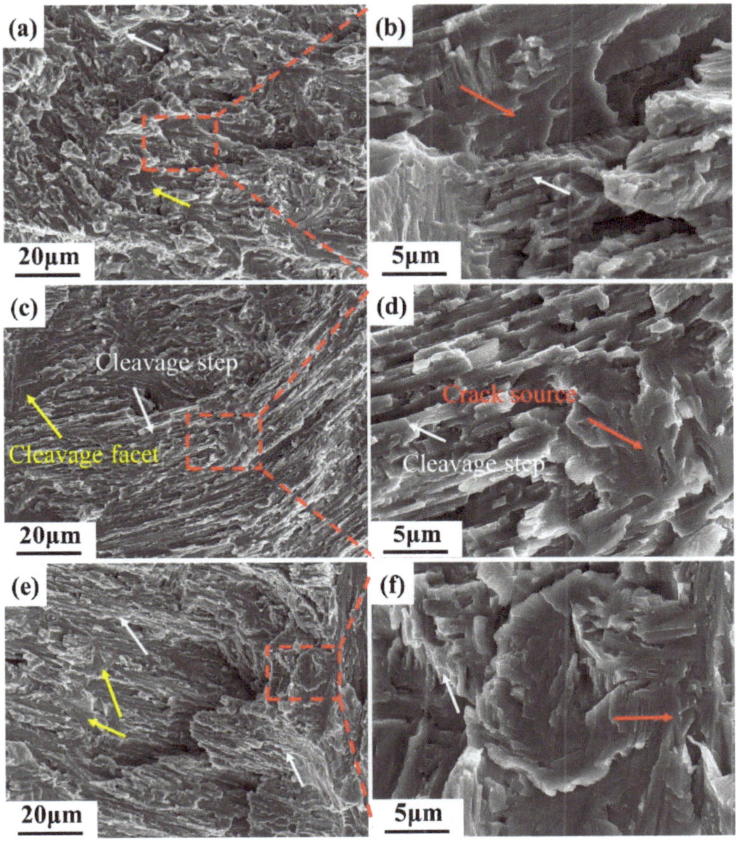

Figure 8. SEM results of fracture surfaces of Ti4822-xNb alloys (yellow arrows indicate cleavage facets; white arrows indicated cleavage steps; red arrows indicated the source of the cracks): (**a**) S1, (**b**) higher magnification image of the region in the red box in (**a**), (**c**) S2, (**d**) higher magnification image of the region in the red box in (**c**), (**e**) S3, and (**f**) higher magnification image of the region in the red box in (**e**).

3.4. Tribological Properties

Figure 9a displays curves for the coefficient of friction for the alloys during wear testing in air under a 10 N normal load. The coefficient of friction initially increased with increasing sliding time and stabilized after ~5 min. In the initial run-in period, the actual contact area was small, resulting in a low friction coefficient [56]. After a period of running, the contact area gradually increased as materials wore away. During the stable stage, the friction coefficients of the three samples still fluctuated somewhat, but the fluctuations decreased with time. Figure 9b shows the average friction coefficients for S1, S2, and S3 as 0.55 ± 0.05, 0.58 ± 0.06, and 0.61 ± 0.08, respectively. The errors of the friction coefficient reflect the degree of fluctuation. Both the friction coefficient and the fluctuations slightly increased as the Nb content rose. The wear volume, V, was calculated from the following:

$$V = 2\pi R \cdot S \tag{2}$$

where R is the radius of the test and S is the cross-sectional area of the wear track. The radius of the abrasion paths was measured, and the values of S were determined using the ultra-depth-of-field three-dimensional microscopic system, as shown in Figure 9. The resulting wear volumes for S1, S2, and S3 were 60.8 ± 0.1 mm^3, 72.5 ± 2.6 mm^3, and

76.2 ± 0.6 mm^3, respectively, see Figure 9c, i.e., the wear volume increased as the Nb content increased. Figure 10 clearly shows that there was a pileup of material at the boundary between the track and the unworn area, indicating the existence of local plastic deformation in all the alloys. It has been demonstrated that, in the absence of any surface irregularities, the maximum shear stress under the surface of a rigid sphere pressed against a flat material can be up to 0.48 times the contact stress [57]. However, in practice, surface roughness necessitates considering additional factors, resulting in a three times higher accurate shear stress level [58]. When a normal load of 10 N is applied, the normal surface contact stress is approximately 1000 MPa, corresponding to a sub-surface shear stress of over 1400 MPa. This is considerably higher than the yield strength of Ti4822-xNb alloys [59]. As a result, plastic deformation occurs during wear testing, leading to a pile-up at the edge of the wear track.

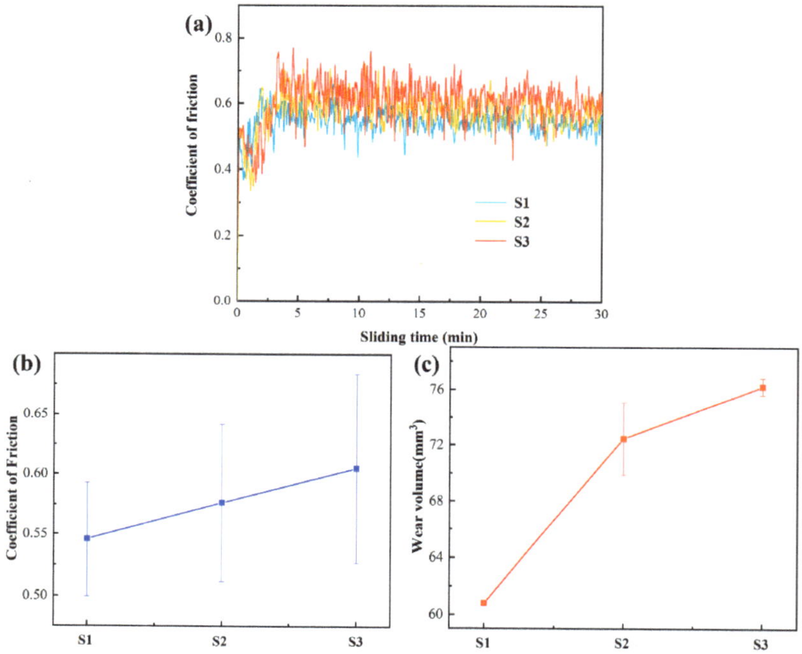

Figure 9. (**a**) Friction coefficient-sliding time curves, (**b**) average friction coefficients, and (**c**) wear volumes for the Ti4822-xNb alloys.

To analyze the wear mechanisms of the alloys, the wear tracks of samples were examined further. Figure 11 shows secondary electron images. The wear surfaces of the three samples were similar, and the wear tracks were generally relatively flat. The surface of the wear tracks contained a significant amount of white and bright particles, which were wear debris [60]. The surface damage consisted of tens of micrometers wide and long continuous plowed grooves. This indicates that the primary wear mechanism of TiAl alloy is abrasive wear, which is consistent with reports in the literature [7,61,62]. Upon further magnification, gray "scars" (the yellow arrows in Figure 11b,e,h) could been seen were present on the wear surfaces, which were clearly indications of plastic deformation. Figure 11c,f,i shows the scars in detail. In the S2 sample, three typical morphologies were analyzed: wear debris, matrix, and scar. Point analyses were carried out at A, B, and C in Figure 11, and the results are presented in Table 3. The results revealed that wear debris showed a high O content of 31.7%, which indicates that oxidative wear occurred. The matrix contained only a small amount of O, suggesting that no tribo-oxide layer formed on the wear surfaces. The O content at the scar was between matrix and wear debris. All three sites contained only trace amounts of Si.

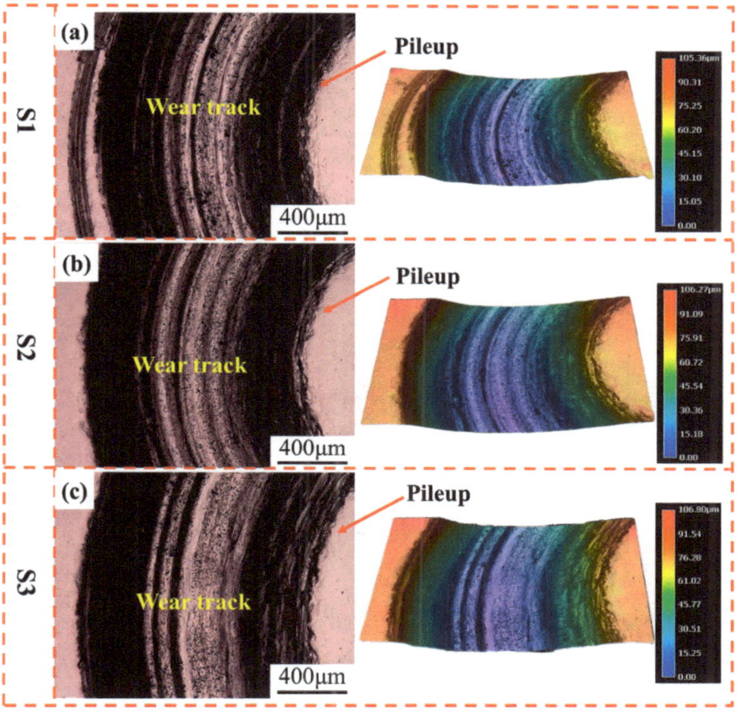

Figure 10. Surface observation and 3D images from ultra-depth three-dimensional microscope for wear tracks of Ti4822-xNb alloys: (**a**) S1, (**b**) S2, and (**c**) S3.

As previously stated, the surface of the samples was subjected to considerably elevated stresses compared to its yield strength during friction, which resulted in plastic deformation of the asperities (Figure 11c,f,i). Upon reaching a certain threshold, plastic deformation ceded to the generation of wear debris. Some of the wear debris was expelled from the wear testing system under the force of centrifugal acceleration. At the same time, the rest remained between the wearing components, continuing to contribute to the wear process. The heat generation by friction resulted in an elevated local temperature, which stimulated the fine abrasive particles to react with oxygen in the air to form metal oxide particles (Table 3). The action of both frictional and normal stresses led to the agglomeration, compaction, and welding of these metal oxides, as shown in Figure 12. Nevertheless, a friction oxide layer with good wear resistance was not formed further. Consequently, the wear process became more intense, and the coefficient of friction exhibited considerable fluctuations.

Table 3. EDS results from the marked areas in Figure 11e.

Label	Composition in at.%					
	Ti	Al	Nb	Cr	O	Si
A	31.4	32.7	2.8	1.2	31.7	0.1
B	46.9	43.1	3.5	2.0	4.3	0.2
C	40.7	39.1	3.2	1.7	15.2	0.1

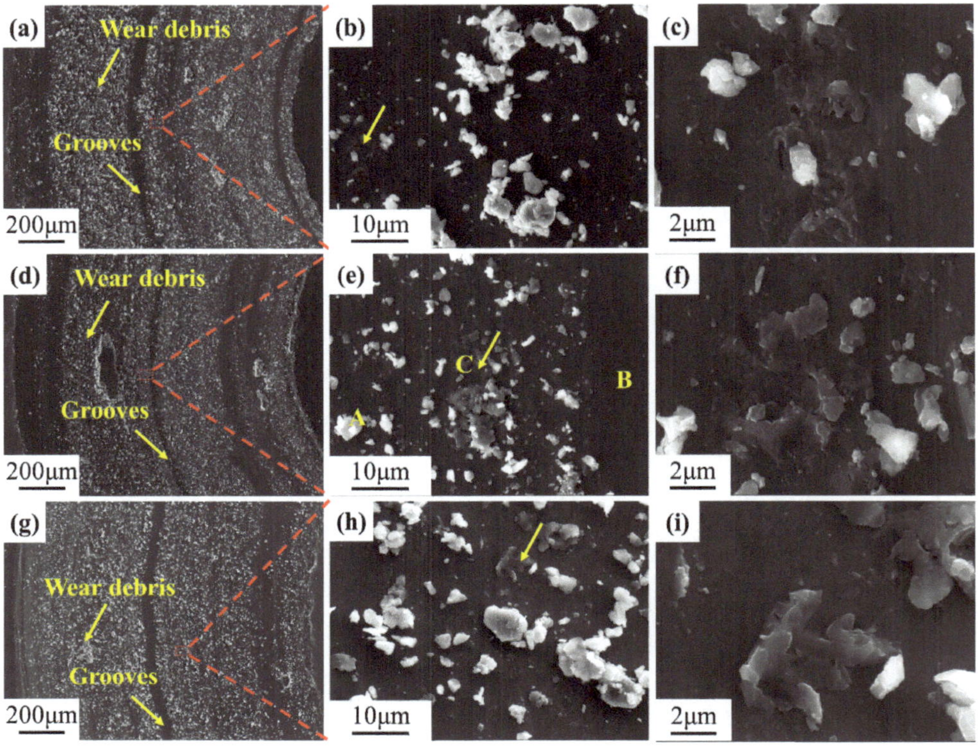

Figure 11. Secondary electron images showing the wear surface of Ti4822-xNb alloys and corresponding EDS maps: (**a**) S1, (**b**) a higher magnification image of the region in the red box in (**a**), (**c**) a higher magnification image of the region indicated by the yellow arrow in (**b**), (**d**) S2, (**e**) a higher magnification image of the region in the red box in (**d**) (the yellow letters indicate different characteristic areas), (**f**) a higher magnification image of the region indicated by the yellow arrow in (**e**), (**g**) S3, (**h**) a higher magnification image of the region in the red box in (**g**), (**i**) a higher magnification image of the region indicated by the yellow arrow in (**h**).

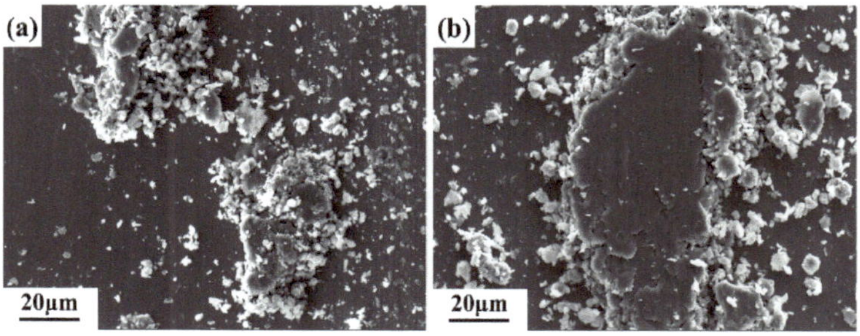

Figure 12. Secondary electron images showing (**a**) agglomeration and (**b**) compaction of wear debris.

In the presence of a load, the Si_3N_4 ceramic ball was capable of exerting pressure on the surface of the alloy samples, resulting in the material undergoing flow. In this context, the role of shear stress was of particular significance. It has been demonstrated that the γ/γ and α_2/γ interfaces in TiAl alloys exert a significant blocking effect on the sliding of the anti-friction material [46]. In comparison to S1, S2 and S3 contained a considerable number of lamellar interfaces. Further, the increased Nb content resulted in a more severe lattice distortion of the γ phase, necessitating a higher shear stress for material flow. This manifested itself macroscopically as an increase in the coefficient of friction. However, the wear volume of the three samples showed an opposite trend. The microstructure is the primary factor influencing the tribological behavior under identical sliding conditions. S1 exhibited a microstructure with more β_0 phases (5.7%) than the other two alloys (1–1.9%) and thus demonstrated better plasticity and toughness under local high-temperature conditions, which may reduce the wear and plowing action of Si_3N_4 and oxide particles. Furthermore, the increase in Nb content resulted in a more pronounced S-segregation in the alloys, accompanied by a reduction in microstructural homogeneity. As previously stated, the lamellar colonies of S3 were smaller and more numerous than those of S2. Additionally, delicate γ grain clusters were present at the boundaries of the lamellar colonies of both. It has been reported in the literature that the hardness of lamellar structures in TiAl alloys is higher than that of massive or equiaxed γ grains [28]. The hardness of a material reflects the ability to resist deformation or damage to its surface. When subjected to shear stress, the uneven distribution of strain between structures with different hardness values leads to the initiation of cracks. The morphologies of the abrasive debris in all three samples can be observed in Figure 13. As the Nb content increased, the size of abrasive debris also increased, particularly in S3. The heterogeneous composition and structure of the alloy rendered it susceptible to the detachment of large pieces during friction, which in turn resulted in an augmented rate of wear. This is consistent with the preceding analysis. Therefore, the wear performance of the alloy diminished with the increase in Nb content.

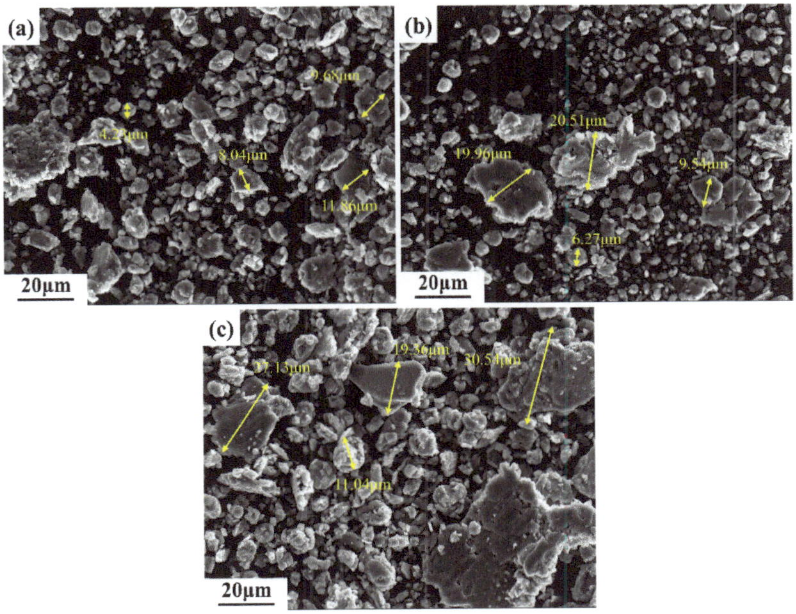

Figure 13. Secondary electron images showing the wear debris from (**a**) S1, (**b**) S2, and (**c**) S3.

4. Conclusions

In this work, Ti4822 alloys with different additional Nb contents were successfully prepared using LMD processing. The microstructure, mechanical, and tribological properties of the as-fabricated TiAl alloys were characterized and analyzed. The following conclusions can be drawn:

1. The TiAl alloys prepared by LMD showed a decrease in the β_0 phase and an increase in the γ phase with increasing Nb content. The presence of the residual β_0 phase was primarily caused by Cr segregation and weakly correlated with the Nb content. Conversely, the increase in Nb altered the solidification path of TiAl alloys from

peritectic solidification to β-solidification. The increase in Nb content obviously refined the microstructures and increased the tendency for S-segregation.

2. As the Nb content increased, the microhardness of the sample initially increased and then decreased with the alloy containing an additional 2 at.% Nb exhibiting the highest hardness value of 359.2 ± 6.5 $HV_{0.3}$. Furthermore, this alloy exhibited a fracture strength of 568 ± 7.8 MPa, which was nearly 100 MPa higher than that of the modified alloy with an additional 1 at. % Nb. The alloy with more Nb (4 at.%) showed the same strength as that with an additional 2 at.%.
3. The wear resistance of the as-fabricated alloys decreased with increasing Nb content. The main wear mechanism for all alloys was abrasive wear, while oxidative wear occurred without the formation of tribo-oxide layers. In addition, local plastic deformation and pile-ups at the edges of the wear tracks occurred.

Author Contributions: K.H., conceptualization, data curation, methodology, and writing—original draft. F.X., conceptualization, formal analysis, investigation, and methodology. X.L., conceptualization and writing—review and editing. S.L., conceptualization. Q.W., writing—review and editing, methodology. I.B., conceptualization and methodology. M.S., conceptualization and methodology. H.W., funding acquisition and writing—review and editing. All authors have read and agreed to the published version of the manuscript.

Funding: This research was funded by the National Natural Science Foundation of China (No. 12374022, U23A20540), the Natural Science Foundation of Hunan Province for Distinguished Young Scholars (No. 2023JJ10075), the Scientific and Technological Project of Yunnan Precious Metals Laboratory (No. YPML-202305247), and the Central South University Research Program of Advanced Interdisciplinary Studies (No. 2023QYJC038).

Institutional Review Board Statement: Not applicable.

Informed Consent Statement: Not applicable.

Data Availability Statement: The data presented in this study are available on request from the corresponding author.

Acknowledgments: We thank the electron microscope center of KAIPLE Co., Ltd. (Changsha, China) for the support of microstructural characterizations.

Conflicts of Interest: Authors Feng Xu and Xinyan Liu were employed by the company Farsoon Technologies Co., Ltd. The remaining authors declare that the research was conducted in the absence of any commercial or financial relationships that could be construed as a potential conflict of interest.

References

1. Clemens, H.; Mayer, S. Design, Processing, Microstructure, Properties, and Applications of Advanced Intermetallic TiAl Alloys. *Adv. Eng. Mater.* **2013**, *15*, 191–215. [CrossRef]
2. Lu, Q.; Xu, Z.; Chen, X.; Feng, P.; Li, C.; Yi, J. Evading the strength and ductility trade-off dilemma in titanium matrix composites through supersaturated solid solution and phase transformation. *Mater. Sci. Eng. A* **2024**, *894*, 146211. [CrossRef]
3. Park, J.-S.; Yang, G.; Kim, S.-W. A high tensile strength above 900 °C in β-solidified TiAl alloy through alloy design and microstructure optimization. *J. Alloys Compd.* **2023**, *947*, 169676. [CrossRef]
4. Zheng, G.; Tang, B.; Zhao, S.; Wang, J.; Xie, Y.; Chen, X.; Wang, W.Y.; Liu, D.; Yang, R.; Li, J. Breaking the high-temperature strength-ductility trade-off in TiAl alloys through microstructural optimization. *Int. J. Plast.* **2023**, *170*, 103756. [CrossRef]
5. Gao, P.; Huang, W.; Yang, H.; Jing, G.; Liu, Q.; Wang, G.; Wang, Z.; Zeng, X. Cracking behavior and control of β-solidifying Ti-40Al-9V-0.5Y alloy produced by selective laser melting. *J. Mater. Sci. Technol.* **2020**, *39*, 144–154. [CrossRef]
6. Cho, K.; Kobayashi, R.; Oh, J.Y.; Yasuda, H.Y.; Todai, M.; Nakano, T.; Ikeda, A.; Ueda, M.; Takeyama, M. Influence of unique layered microstructure on fatigue properties of Ti-48Al-2Cr-2Nb alloys fabricated by electron beam melting. *Intermetallics* **2018**, *95*, 1–10. [CrossRef]
7. Lu, X.; Li, J.; Chen, X.; Qiu, J.; Wang, Y.; Liu, B.; Liu, Y.; Rashad, M.; Pan, F. Mechanical, tribological and electrochemical corrosion properties of in-situ synthesized Al_2O_3/TiAl composites. *Intermetallics* **2020**, *120*, 106758. [CrossRef]
8. Liu, S.; Ding, H.; Chen, R.; Guo, J.; Fu, H. Evolution of rapidly grown cellular microstructure during heat treatment of TiAl-based intermetallic and its effect on micromechanical properties. *Intermetallics* **2021**, *132*, 107166. [CrossRef]
9. Wang, J.; Pan, Z.; Wei, L.; He, S.; Cuiuri, D.; Li, H. Introduction of ternary alloying element in wire arc additive manufacturing of titanium aluminide intermetallic. *Addit. Manuf.* **2019**, *27*, 236–245. [CrossRef]

10. Liang, Y.-F.; Xu, X.-J.; Lin, J.-P. Advances in phase relationship for high Nb-containing TiAl alloys. *Rare Met.* **2016**, *35*, 15–25. [CrossRef]
11. Duan, B.; Yang, Y.; He, S.; Feng, Q.; Mao, L.; Zhang, X.; Jiao, L.; Lu, X.; Chen, G.; Li, C. History and development of γ-TiAl alloys and the effect of alloying elements on their phase transformations. *J. Alloys Compd.* **2022**, *909*, 164811. [CrossRef]
12. Yu, J.; Lin, X.; Wang, J.; Chen, J.; Huang, W. Mechanics and energy analysis on molten pool spreading during laser solid forming. *Appl. Surf. Sci.* **2010**, *256*, 4612–4620. [CrossRef]
13. Schwaighofer, E.; Clemens, H.; Mayer, S.; Lindemann, J.; Klose, J.; Smarsly, W.; Güther, V. Microstructural design and mechanical properties of a cast and heat-treated intermetallic multi-phase γ-TiAl based alloy. *Intermetallics* **2014**, *44*, 128–140. [CrossRef]
14. Luo, Y.; Liu, S.; Sun, Z.; Liu, B.; Wang, L.; Wang, Y.; Liu, Y. Microstructural evolution during annealing of a powder metallurgical TiAl-Nb composite and its effect on mechanical properties. *J. Mater. Res. Technol.* **2023**, *25*, 3654–3669. [CrossRef]
15. Schloffer, M.; Iqbal, F.; Gabrisch, H.; Schwaighofer, E.; Schimansky, F.-P.; Mayer, S.; Stark, A.; Lippmann, T.; Göken, M.; Pyczak, F.; et al. Microstructure development and hardness of a powder metallurgical multi phase γ-TiAl based alloy. *Intermetallics* **2012**, *22*, 231–240. [CrossRef]
16. Zhang, S.; Tian, N.; Li, J.; Yang, G.; Yang, W.; Wang, G.; Liu, Z.; Li, Y. Microstructure evolution of a forged TiAl-Nb alloy during high-temperature tensile testing. *Mater. Charact.* **2023**, *205*, 113268. [CrossRef]
17. Zhang, S.Z.; Zhao, Y.B.; Zhang, C.J.; Han, J.C.; Sun, M.J.; Xu, M. The microstructure, mechanical properties, and oxidation behavior of beta-gamma TiAl alloy with excellent hot workability. *Mater. Sci. Eng. A* **2017**, *700*, 366–373. [CrossRef]
18. Liu, S.; Ding, H.; Zhang, H.; Chen, R.; Guo, J.; Fu, H. High-density deformation nanotwin induced significant improvement in the plasticity of polycrystalline γ-TiAl-based intermetallic alloys. *Nanoscale* **2018**, *10*, 11365–11374. [CrossRef]
19. Li, H.; Qi, Y.; Liang, X.; Zhu, Z.; Lv, F.; Liu, Y.; Yang, Y. Microstructure and high temperature mechanical properties of powder metallurgical Ti-45Al-7Nb-0.3W alloy sheets. *Mater. Des.* **2016**, *106*, 90–97. [CrossRef]
20. Ren, Y.; Han, B.; Wu, H.; Wang, J.; Liu, B.; Wei, B.; Jiao, Z.; Baker, I. Copper segregation-mediated formation of nanotwins and 9R phase in titanium alloys produced by laser powder bed fusion. *Scr. Mater.* **2023**, *224*, 115115. [CrossRef]
21. Zhang, T.; Liu, C.-T. Design of titanium alloys by additive manufacturing: A critical review. *Adv. Powder Mater.* **2022**, *1*, 100014. [CrossRef]
22. Xue, H.; Liu, C.; Song, Y.; Liang, Y.; Tong, X.; Wang, Y.; Lin, J. Additive manufacturing of nano-W composite high Nb-TiAl alloys fabricated via selective laser melting. *Mater. Lett.* **2023**, *347*, 134569. [CrossRef]
23. Liu, M.; Wang, J.; Hu, T.; Xu, S.; Shuai, S.; Xuan, W.; Yin, S.; Chen, C.; Ren, Z. Laser powder bed fusion of a Ni3Al-based intermetallic alloy with tailored microstructure and superior mechanical performance. *Adv. Powder Mater.* **2024**, *3*, 100152. [CrossRef]
24. Sun, J.e.; Zhang, B.; Qu, X. High strength Al alloy development for laser powder bed fusion. *J. Micromech. Mol. Phys.* **2021**, *6*, 2141001. [CrossRef]
25. DebRoy, T.; Wei, H.L.; Zuback, J.S.; Mukherjee, T.; Elmer, J.W.; Milewski, J.O.; Beese, A.M.; Wilson-Heid, A.; De, A.; Zhang, W. Additive manufacturing of metallic components—Process, structure and properties. *Prog. Mater Sci.* **2018**, *92*, 112–224.
26. Xue, H.; Liang, Y.; Peng, H.; Wang, Y.; Shang, S.-L.; Liu, Z.-K.; Lin, J. An additively manufactured γ-based high Nb-TiAl composite via coherent interface regulation. *Scr. Mater.* **2023**, *223*, 115102. [CrossRef]
27. Zhu, Y.; Wang, Z.; Yu, B.; Li, G.; Xue, Y.; Liang, Y.-J. Additive manufacturing of fine-grain fully lamellar titanium aluminide alloys. *Mater. Des.* **2023**, *230*, 111989. [CrossRef]
28. Wu, Y.; Zhang, S.; Cheng, X.; Wang, H. Investigation on solid-state phase transformation in a Ti-47Al-2Cr-2V alloy due to thermal cycling during laser additive manufacturing process. *J. Alloys Compd.* **2019**, *799*, 325–333. [CrossRef]
29. Wang, D.; Yang, Y.; Wang, Y.; Yang, L.; Wang, H.; Yang, S. Introduction to the Special Issue on Design and Simulation in Additive Manufacturing. *Cmes-Comput. Model. Eng. Sci.* **2021**, *126*, 1–4.
30. Cao, L. Mesoscopic-Scale Numerical Investigation Including the Influence of Process Parameters on LPBF Multi-Layer Multi-Path Formation. *Cmes-Comput. Model. Eng. Sci.* **2021**, *126*, 5–23.
31. Liu, Z.; Zhao, D.; Wang, P.; Yan, M.; Yang, C.; Chen, Z.; Lu, J.; Lu, Z. Additive manufacturing of metals: Microstructure evolution and multistage control. *J. Mater. Sci. Technol.* **2022**, *100*, 224–236. [CrossRef]
32. Zhang, X.; Li, C.; Wu, M.; Ye, Z.; Wang, Q.; Gu, J. Atypical pathways for lamellar and twinning transformations in rapidly solidified TiAl alloy. *Acta Mater.* **2022**, *227*, 117718. [CrossRef]
33. Sui, S.; Chew, Y.; Hao, Z.; Weng, F.; Tan, C.; Du, Z.; Bi, G. Effect of cyclic heat treatment on microstructure and mechanical properties of laser aided additive manufacturing Ti–6Al–2Sn–4Zr–2Mo alloy. *Adv. Powder Mater.* **2022**, *1*, 100002. [CrossRef]
34. Fan, H.; Liu, Y.; Yang, S. Martensite decomposition during post-heat treatments and the aging response of near-α Ti–6Al–2Sn–4Zr–2Mo (Ti-6242) titanium alloy processed by selective laser melting (SLM). *J. Micromech. Mol. Phys.* **2021**, *6*, 2050018. [CrossRef]
35. Sharman, A.; Hughes, J.; Ridgway, K. Characterisation of titanium aluminide components manufactured by laser metal deposition. *Intermetallics* **2018**, *93*, 89–92. [CrossRef]
36. Zheng, G.; Tang, B.; Chen, W.; Zhao, S.; Xie, Y.; Chen, X.; Li, J.; Zhu, L. Long-period stacking ordering induced ductility of nanolamellar TiAl alloy at elevated temperature. *Mater. Res. Lett.* **2023**, *11*, 414–421. [CrossRef]

37. Wang, J.; Luo, Q.; Wang, H.; Wu, Y.; Cheng, X.; Tang, H. Microstructure characteristics and failure mechanisms of Ti-48Al-2Nb-2Cr titanium aluminide intermetallic alloy fabricated by directed energy deposition technique. *Addit. Manuf.* **2020**, *32*, 10[illegible]7. [CrossRef]
38. Zhang, X.; Li, C.; Zheng, M.; Ye, Z.; Yang, X.; Gu, J. Anisotropic tensile behavior of Ti-47Al-2Cr-2Nb alloy fabricated by direct laser deposition. *Addit. Manuf.* **2020**, *32*, 101087. [CrossRef]
39. Fatoba, O.S.; Lasisi, A.M.; Ikumapayi, O.M.; Akinlabi, S.A.; Akinlabi, E.T. Computational modelling of laser additive manufactured (LAM) Titanium alloy grade 5. *Mater. Today Proc.* **2021**, *44*, 1254–1262. [CrossRef]
40. Shi, X.; Wang, H.; Feng, W.; Zhang, Y.; Ma, S.; Wei, J. The crack and pore formation mechanism of Ti–47Al–2Cr–2Nb alloy fabricated by selective laser melting. *Int. J. Refract. Met. Hard Mater.* **2020**, *91*, 105247. [CrossRef]
41. Chlupová, A.; Heczko, M.; Obrtlík, K.; Polák, J.; Roupcová, P.; Beran, P.; Kruml, T. Mechanical properties of high niobium TiAl alloys doped with Mo and C. *Mater. Des.* **2016**, *99*, 284–292. [CrossRef]
42. Xu, R.; Li, M.; Zhao, Y. A review of microstructure control and mechanical performance optimization of γ-TiAl alloys. *J. Alloys Compd.* **2023**, *932*, 167611. [CrossRef]
43. Tetsui, T. Effects of high niobium addition on the mechanical properties and high-temperature deformability of gamma TiAl alloy. *Intermetallics* **2002**, *10*, 239–245. [CrossRef]
44. Liu, Z.C.; Lin, J.P.; Li, S.J.; Chen, G.L. Effects of Nb and Al on the microstructures and mechanical properties of high Nb containing TiAl base alloys. *Intermetallics* **2002**, *10*, 653–659. [CrossRef]
45. Fang, H.; Chen, R.; Liu, Y.; Tan, Y.; Su, Y.; Ding, H.; Guo, J. Effects of niobium on phase composition and improving mechanical properties in TiAl alloy reinforced by Ti_2AlC. *Intermetallics* **2019**, *115*, 106630. [CrossRef]
46. Wang, H.; Wang, Q.; Zeng, L.; Zhang, H.; Ding, H. Microstructure, mechanical and tribological performances of a directionally solidified γ-TiAl alloy. *Mater. Charact.* **2021**, *179*, 111393. [CrossRef]
47. Cheng, J.; Yang, J.; Zhang, X.; Zhong, H.; Ma, J.; Li, F.; Fu, L.; Bi, Q.; Li, J.; Liu, W. High temperature tribological behavior of a Ti-46Al-2Cr-2Nb intermetallics. *Intermetallics* **2012**, *31*, 120–126. [CrossRef]
48. Wang, Q.; Ding, H.; Zhang, H.; Chen, R.; Guo, J.; Fu, H. Influence of Mn addition on the microstructure and mechanical properties of a directionally solidified γ-TiAl alloy. *Mater. Charact.* **2018**, *137*, 133–141. [CrossRef]
49. Chen, G.L.; Xu, X.J.; Teng, Z.K.; Wang, Y.L.; Lin, J.P. Microsegregation in high Nb containing TiAl alloy ingots beyond laboratory scale. *Intermetallics* **2007**, *15*, 625–631. [CrossRef]
50. Zhang, W.J.; Liu, Z.C.; Chen, G.L.; Kim, Y.W. Deformation mechanisms in a high-Nb containing γ–TiAl alloy at 900 °C. *Mater. Sci. Eng. A* **1999**, *271*, 416–423. [CrossRef]
51. Nath, P.; Bar, H.N.; Bhattacharjee, A.; Sen, I. Designing of novel microstructure and its impact on the improved service temperature mechanical performance of 2nd and 3rd generation advanced intermetallic TiAl alloys. *Mater. Sci. Eng. A* **2024**, *893*, 146[illegible]. [CrossRef]
52. Wang, Q.; Liu, X.; Ren, Y.; Song, M.; Baker, I.; Wu, H. Microstructural evolution and cryogenic and ambient temperature deformation behavior of the near-α titanium alloy TA15 fabricated by laser powder bed fusion. *J. Alloys Compd.* **2024**, *1001*, 175[illegible]75. [CrossRef]
53. He, P.; Webster, R.F.; Yakubov, V.; Kong, H.; Yang, Q.; Huang, S.; Ferry, M.; Kruzic, J.J.; Li, X. Fatigue and dynamic aging behavior of a high strength Al-5024 alloy fabricated by laser powder bed fusion additive manufacturing. *Acta Mater.* **2021**, *220*, 117[illegible]. [CrossRef]
54. Singh, V.; Mondal, C.; Sarkar, R.; Bhattacharjee, P.P.; Ghosal, P. Effects of Cr alloying on the evolution of solidification microstructure and phase transformations of high-Nb containing γ-TiAl based alloys. *Intermetallics* **2021**, *131*, 107117. [CrossRef]
55. Fang, H.; Chen, R.; Chen, X.; Yang, Y.; Su, Y.; Ding, H.; Guo, J. Effect of Ta element on microstructure formation and mechanical properties of high-Nb TiAl alloys. *Intermetallics* **2019**, *104*, 43–51. [CrossRef]
56. Agbedor, S.-O.; Wu, H.; Ren, Y.; Liang, L.; Yang, D.; Liu, B.; Liu, Y.; Baker, I. A two-decade odyssey in fusion-based additive manufacturing of titanium alloys and composites. *Appl. Mater. Today* **2024**, *39*, 102242. [CrossRef]
57. Ovcharenko, A.; Halperin, G.; Verberne, G.; Etsion, I. In situ investigation of the contact area in elastic-plastic spherical contact during loading-unloading. *Tribol. Lett.* **2007**, *25*, 153–160. [CrossRef]
58. Rastkar, A.R.; Bloyce, A.; Bell, T. Sliding wear behaviour of two gamma-based titanium aluminides. *Wear* **2000**, *240*, 19–26. [CrossRef]
59. Wang, Q.; Ding, H.; Zhang, H.; Chen, R.; Guo, J.; Fu, H. Variations of microstructure and tensile property of γ-TiAl alloys with 0–0.5 at% C additives. *Mater. Sci. Eng. A-Struct. Mater. Prop. Microstruct. Process.* **2017**, *700*, 198–208. [CrossRef]
60. Du, J.; Ren, Y.; Liu, X.; Xu, F.; Wang, X.; Zhou, R.; Baker, I.; Wu, H. Microstructural Evolution, Mechanical Properties and Tribological Behavior of B(4)C-Reinforced Ti In Situ Composites Produced by Laser Powder Bed Fusion. *Materials* **2023**, *16*, 4[illegible]. [CrossRef]

61. Cheng, F.; Lin, J.; Liang, Y. Friction and wear properties of a high Nb-containing TiAl alloy against WC-8Co, Si_3N_4, and GCr15 in an unlubricated contact. *Intermetallics* **2019**, *106*, 7–12. [CrossRef]
62. Zhou, H.; Su, Y.; Liu, N.; Kong, F.; Wang, X.; Zhang, X.; Chen, Y. Modification of microstructure and properties of Ti-47Al-2Cr-4Nb-0.3W alloys fabricated by SPS with trace multilayer graphene addition. *Mater. Charact.* **2018**, *138*, 1–10. [CrossRef]

materials

Article

Tailored Time–Temperature Transformation Diagram for IN718 Alloy Obtained via Powder Bed Fusion Additive Manufacturing: Phase Behavior and Precipitation Dynamic

Julio Cesar Franco-Correa [1,2], Enrique Martínez-Franco [2], Celso Eduardo Cruz-González [2], Juan Manuel Salgado-López [2] and Jhon Alexander Villada-Villalobos [2,3,*]

1 Posgrado Interinstitucional en Ciencia y Tecnología PICYT, CIDESI, Queretaro 76125, Mexico; jczfrancoc@gmail.com
2 Center for Engineering and Industrial Development, CIDESI, Av. Pie de la Cuesta 702, Santiago de Querétaro 76125, Mexico; enrique.martinez@cidesi.edu.mx (E.M.-F.); ecruz@cidesi.edu.mx (C.E.C.-G.); msalgado@cidesi.edu.mx (J.M.S.-L.)
3 Consejo Nacional de Humanidades, Ciencia y Tecnología (CONAHCYT), Av. Insurgentes Sur 1582, Col. Crédito Constructor, Demarcación Territorial Benito Juárez, Mexico City 03940, Mexico
* Correspondence: jhon.villada@cidesi.edu.mx

Abstract: Experimental and computational approaches were used to study the microstructure of IN718 produced via powder bed fusion additive manufacturing (PBF-AM). The presence, chemical composition, and distribution of stable and metastable phases (γ'', δ, MC, and Laves) were also analyzed. The information obtained from the microstructural study was used to construct a tailored time–temperature transformation (TTT) diagram customized for additive manufacturing of IN718. Experimental techniques, including differential scanning calorimetry (DSC), scanning electron microscopy, energy dispersive X-ray spectroscopy, and electron backscatter diffraction (EBSD), were employed to establish the morphological, chemical, and structural characteristics of the microstructure. The Thermo-Calc software and a Scheil–Gulliver model were used to analyze the presence and behavior of phase transformations during heating and cooling processes under non-thermodynamic equilibrium conditions, typical of AM processes. Unlike conventional TTT diagrams of this alloy, the diagram presented here reveals that the precipitation of γ'' and δ phases occurs at lower temperatures and shorter times in AM-manufactured parts. Significantly, the superposition of γ'' and δ phase curves in the proposed diagram underscores the interdependence between these phases. This TTT diagram is a valuable insight that can help in the development of heat treatment processes and quality control for IN718 produced via PBF-AM.

Keywords: IN718; stable and metastable phases; powder bed fusion (PBF); thermodynamic behavior; additive manufacturing (AM); phase transformations

1. Introduction

Additive manufacturing (AM) introduces new production methods and allows customization of materials' geometric shape, chemical composition, and microstructure, enhancing the performance of functionally graded materials (FGM) [1]. In recent years, researchers such as [2–4] have conducted experimental studies on phase transformations, melting and solidification structures, residual stress, and optimal processing parameters, revealing essential information on surface topography, grain structure, crystalline structure, phases, and precipitates. These studies have acquired correlated thermodynamics and kinetics behavior, diffusion phenomena, and thermal history in order to predict phase transformations in alloys obtained via AM. On the other hand, simulations in AM offer a powerful approach for extracting valuable information about material properties. Using thermodynamic modeling, such as CALPHAD methodology, it becomes feasible to predict equilibrium phases in a multi-component system [5].

IN718 is a nickel-based superalloy well known for its exceptional mechanical properties [6]. These exceptional properties are achieved via a precipitation hardening mechanism through the use of AMS5662 [7] and AMS5383 [8] heat treatments suitable for components manufactured using conventional methods such as forging or casting [9]. These treatments promote the precipitation of γ' and γ'' phases, which provides the material with exceptional properties of high resistance to stress, fatigue, rupture, and creep at high temperatures. However, in AM, the microstructural behavior of IN718 is influenced by several factors, including solidification under non-thermodynamic equilibrium conditions, high-temperature gradients, and thermal history during fabrication.

As a result, precipitation, solidification, and reinforcing phase growth mechanisms can exhibit notable differences compared to traditional manufacturing methods. In their study, Sanchez et al. [10] show a comprehensive summary of the factors impacting the functional performance of nickel-based superalloys specifically manufactured using additive processes. Morphological defects and microstructures are present and differ from those obtained using conventional manufacturing techniques. Moreover, microstructural characteristics and mechanisms such as phase precipitation, textures, and grain sizes vary from reported thermodynamic predictions. Gallmeyer et al. [11] report that, during the formation of parts obtained via AM, the precipitation of the δ phase is detrimental and undesirable during the nucleation process of reinforcing phases such as γ' and γ''. Naiyuan Xi et al. [12] established that the configurations of the δ phase have a significant effect on the plastic deformation of grains, where the large δ phase at the grain boundary limits the intergranular deformation significantly, but the large intragranular δ phase may cause strain localization and then lead to rapid failure. In summary, the kind, morphology, and location of the phases in the AM process make it difficult to understand the solidification process of the reinforcing phases and, consequently, its impact on mechanical properties.

In a previous work published by our team [13], it has been demonstrated that the effect of conventional heat treatment [7,8] is different when it is applied to AM-produced parts of IN718. For this reason, we decided to go further in the investigation, looking for a TTT diagram that represents the phase transformation as a function of the time and temperature for AM-produced IN718.

In this paper, experimental and computational approaches were used to study the microstructure of IN718 produced via powder bed fusion additive manufacturing (PBF-AM). The presence, chemical composition, and distribution of stable and metastable phases (γ'', δ, MC, and Laves) were also analyzed. The information obtained from the microstructural study (grain size and the composition, location, and distribution of the phases) was used to construct a tailored time–temperature transformation (TTT) diagram customized for additive manufacturing of IN718. Unlike conventional TTT diagrams for this alloy, the diagram presented here reveals that the precipitation of the γ'' and δ phases occurs at lower temperatures and for shorter times in AM-manufactured parts. Significantly, the superposition of the γ'' and δ phase curves in the proposed diagram underscores the interdependence between these phases. The results could be used as a reference for applying a heat treatment to an AM-produced IN718 alloy.

2. Materials and Methods

Figure 1 shows a typical IN718 sample manufactured using powder bed fusion (PBF) technology with an EOSINT M280 3D printer (manufactured by EOS GmbH, Krailling, Germany) in an inert atmosphere (Argon). The printing parameters were optimized to produce samples 10 mm × 10 mm × 10 mm in size with a layer thickness of 0.040 mm, a hatch distance of 0.110 mm, a scan speed of 960 mm/s, a laser power of 285 W, and a 67° layer rotation. The chemical composition of the powder was analyzed using a Varian SpectrAA 220 Atomic Absorption Spectrometer (from Varian, Alto Palo, CA, USA). The measured composition of the powder was as follows: Ni (50.28%), Fe (Balance), Cr (18.27%), Nb (7.27%), Mo (3.48%), Ti (1.1%), Al (0.45%), and C (0.08%). Thermo-Calc® software (https://thermocalc.com (accessed on 13 November 2023), Stockholm, Sweden) version 2019.1, which uses the CALPHAD method, was utilized to predict the phase equilibria in multi-component systems. The Thermo-Calc Nickel-based superalloys databases (TCNI8, MOBNI4) were used to calculate phase fractions, Scheil–Gulliver solidification, and diagrams in temperatures ranging from 300 to 1400 °C. The simulation parameters for AM-manufactured goods, such as dislocation density alloys or the wetting angle, were adopted from the literature [14,15]. Experimental data on morphology and microstructure used in the simulations were obtained from our previous investigation on IN718 [13].

Figure 1. Sample of IN718 manufactured via PBF.

The metallographic preparation of the as-built bulk material was carried out. First, 50 mg of material was obtained using a cutting disk from Buehler (Lake Bluff, IL, USA). Then, this sample was subjected to thermal analysis using a SETSYS Evolution DTA/DSC instrument (Setaram Instrumentation, Caluire-et-Cuire, Francia) according to ASTM D3418 [16] to measure thermal behavior from 25 °C to 1400 °C. The sample was heated at a rate of 10 °C/min and was soaked for 5 min at 1400 °C to homogenize it; then, it was cooled at the same rate.

After the thermal analysis, the sample was prepared for metallographic characterization. This process involved manual grinding using SiC abrasive paper with grit sizes ranging from 320 to 2000, followed by polishing down to 0.5 μm using an alcohol-based diamond suspension and a VibroMet™ machine (Buehler, IL, USA) for 24 h. A Kallings 2 etching step was then applied according to ASTM E407-07 [17], and any residual scratches and deformed traces were eliminated using an IM4000Plus HITACHI ion milling system (Hitachi, Tokyo, Japan) with an accelerating voltage of 6 kV and a rotation speed of 25 rpm for 40 min.

An SEM SU-8230 HITACHI (Hitachi, Tokyo, Japan) equipped with a Bruker e-Flash HR+ electron backscatter diffraction (EBSD) detector was used to analyze the microstructure, texture, grain size, and distribution at 15 Kv and magnification between ×35.0 K and ×150 K. Using EBSD analysis, high-resolution images were acquired at a magnification of ×700, a step size of 1.0 um, and a resolution of 512 × 416 pixels. A grain confidence index standardization was implemented, with angle tolerance set at 10° to ensure accurate and reliable results. Moreover, a neighbor confidence index correlation was considered, requiring a threshold of ≥0.1 and a neighbor orientation correlation.

3. Results and Discussion

3.1. Phase Dynamics versus Temperature

Figure 2 illustrates the prevailing phases and precipitates observed in the IN718 alloy, obtained from Thermo-Calc® software (https://thermocalc.com (accessed on 13 November 2023)) (Stockholm, Sweden). At temperatures below 400 °C, the presence of Laves phases becomes evident. The formation of these intermetallic phases results from the kinetic process of the PBF manufacturing process, thermal history, and the availability of chemical elements. At temperatures above 400 °C, the alloy can exhibit the coexistence of stable (δ) and metastable (γ'') phases with Ni3Nb stoichiometry. The prevalence of each phase is strongly influenced by the availability of chemical elements and the temperature gradient experienced during the manufacturing process. Temperatures exceeding 725 °C reveal the precipitation of MC carbides, mainly NbC. The growth of this carbide is dependent upon the availability of Nb present in the δ and γ'' phases. At the same time, the presence of available Ni facilitates an increase of the Ni-Fe matrix phase. The impact of these dissolutions becomes evident through observable slope changes in the quantities of the matrix phase (green line), the δ phase (orange line), and the γ'' phase (dashed orange line).

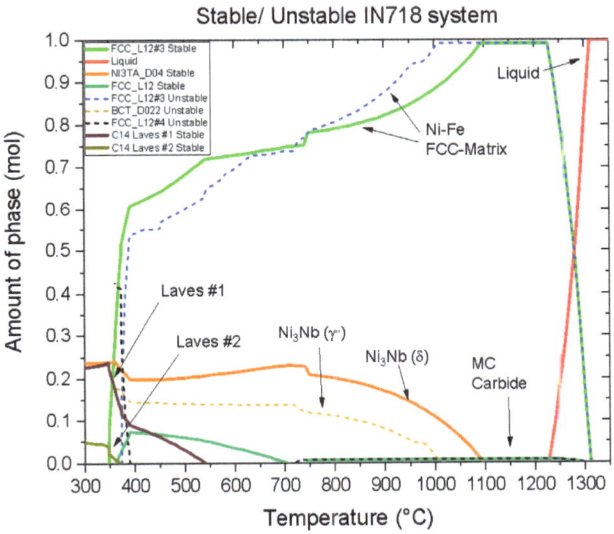

Figure 2. Prediction of phase amounts in the IN718 system for the investigated chemical composition. The graph showcases stable phases (solid lines) and metastable phases (dashed lines).

The presence of stable and metastable phases in IN718 plays a crucial role in determining its mechanical properties. The metastable γ'' phase, in particular, acts as the primary reinforcing precipitate, significantly enhancing the mechanical properties of the alloy. However, over time and with exposure to elevated temperatures during service conditions, this metastable phase transforms into the stable phase δ. This phase transformation leads to changes in the mechanical properties of the material, often resulting in a decrease in strength and potentially affecting other properties.

According to Rayner et al. [18], there is a relationship between DSC temperature peaks and phase transformations. In AM, the thermal characteristics of phase diagrams where there are non-equilibrium thermodynamic conditions cause the formation of metastable phases. Thus, the Scheil–Gulliver solidification model provides an accurate transition temperature of the material, determining heat signals associated with the formation of phases [19].

In Figure 3, the simulation of the solidification profile of the IN718 alloy reveals that the formation of MC carbides starts at a temperature below 1296 °C. In addition, the simulation also supports the preference for the precipitation of NbC, which is in good agreement with experimental findings, exhibiting peaks at 1274 and 1253 °C, as reported by other authors [20,21], where all identifications were based on the solidification pattern Liquid → Liquid + γ → Liquid + γ + NbC → Liquid + γ + NbC + Laves.

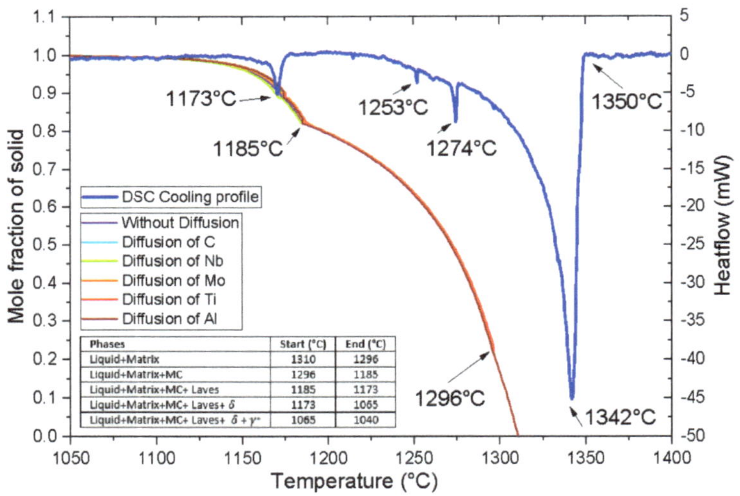

Figure 3. Solidification path under Scheil–Gulliver model (with and without diffusion of alloys) and experimental DSC cooling profile (thick blue line) of IN718 obtained via AM.

Within the temperature range of 1185 to 1173 °C, the formation of Laves phases is distinctly detected. As the temperature falls below 1173 °C, the availability of chemical elements facilitates the stoichiometric formation of the Ni_3Nb phase. However, precisely determining the proportions of γ'' and δ phases in this region remains challenging despite the Scheil–Gulliver model's prediction of the formation of γ''. Particularly, simulations involving diffusive elements show minimal modification to the transformations around 1173 °C. The Scheil–Gulliver model is a widely utilized approach for simulating the solidification behavior of alloys [22]. It is well known that non-equilibrium solidification can induce changes to the solidus temperature due to microsegregation. This phenomenon manifests as variations in composition at the arms of dendrites, and the final part of the liquid solidifies in interdendritic regions with high concentrations of some alloying elements. The model accurately predicts the temperature of phase transformations, but it is important to consider that microsegregation can lead to variations in precipitation temperatures when different diffusive elements are considered. For example, in IN718, low-solubility elements such as Nb, Mo, Ti, and C tend to segregate at the interdendritic regions and form MC carbides and Laves phases [23]. These discrepancies have been observed and reported in several studies, including the research conducted by Knorovsky et al. [24].

The solidification process is a complex phenomenon subject to the influence of various factors, such as the kinetics of solute diffusion, the thermodynamics of microsegregation, and cooling conditions. In the simulation, the diffusion of a single element (Nb, Mo, Al, Ti, and C) was considered, while the other elements followed the classic behavior of the Scheil–Gulliver model. Consequently, it became essential to account for additional factors, including the multicomponent system's kinetics and the manufacturing technique's impact on dendrite geometry and thermal history, to comprehensively understand the process.

Identifying endothermic and exothermic peaks during DSC analysis is crucial in determining critical temperatures, such as the liquidus, eutectic, and nucleation temperatures. In Figure 4, a broader peak emerges during the heating profile near 1305 °C. The onsets of multiple peaks overlap due to solid solution partial melting, as reported by Han et al. [15], resulting in a broad peak with changes in its slope. On the other hand, during the cooling profile near 1342 °C, a broad peak is accompanied by three lower-height peaks: the peaks at 1252 °C and 1274 °C correspond to MC carbides, while the peak at 1171 °C corresponds to the Laves phase, following the path reported by Shi et al. [20]. Heating and cooling enthalpy values of 4.698 mJ/K × mol and 4.596 mJ/K × mol, respectively, were obtained by measuring the area under the curve. This difference is attributed to the exothermic contribution of the three peaks in the total solidification energy associated with the Laves phase and the precipitation of γ'', δ, and MC phases during cooling [25].

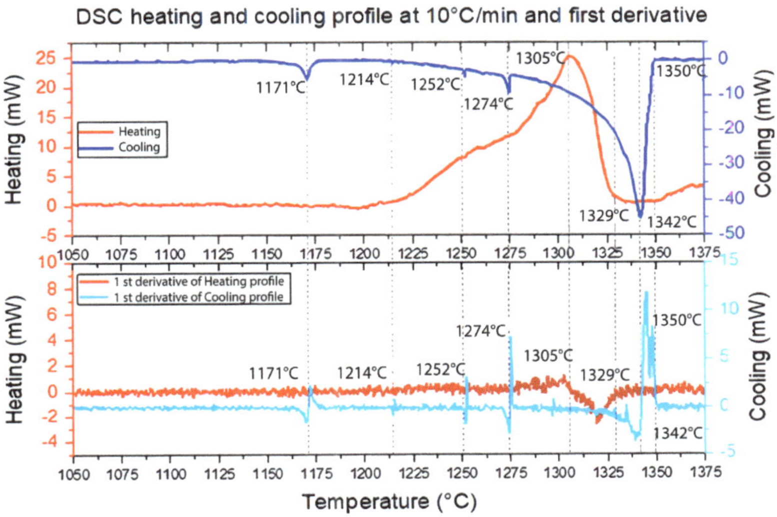

Figure 4. DSC heating and cooling curve profiles at 10 °C/min and the first derivative for IN718 obtained via AM. Several peaks are observed during the cooling process, and a broad peak is observed during the heating. Peak temperatures represent a phase transformation.

3.2. Microstructural Characterization

The manufacturing process conditions deeply influence the microstructure's morphology and crystalline characteristics. As such, it becomes crucial to accurately characterize the microstructural settings to gain a comprehensive understanding of the mechanical properties exhibited by AM-produced samples. Scanning electron microscopy (SEM) and EBSD were employed to identify both morphology and crystallographic orientations within the microstructure.

Figure 5 offers valuable insights into the microstructure, revealing the presence of MC precipitates and Laves phases positioned intragranularly, distinguished by diameters ranging from 0.5 to 2 μm, as reported to occur by Zhao et al. [26] when the high cooling rate during rapid solidification results in the simultaneous formation of Laves phases and MC carbides. Nb and Mo promote Laves formation, while Nb and C promote the precipitation of NbC in interdendritic zones.

The presence of Laves phases in the IN718 alloy significantly impacts its behavior and composition during additive manufacturing, particularly in processes like powder bed fusion. Laves phases are Nb-rich intermetallic compounds. The key consequence of the high concentration of Laves phases is the depletion of Nb in the solid solution, which plays a critical role in the precipitation of the γ'' phase. Since there is insufficient Nb available for γ'' formation, the primary reinforcing phase in this alloy, mechanical properties are notably affected. This depletion of Nb and the consequent alteration in the alloy's microstructure can result in reduced strength, toughness, and other mechanical characteristics.

Carbides are located in different zones of the microstructure, within the grain, and at the grain boundaries. The first ones reduce dislocation movement inside the grains, improving creep resistance and hardness. On the other hand, carbides at the grain boundaries could be of two types: primary (MC) or secondary carbides ($M_{23}C_6$). The primary carbides reduce grain sliding, improving the creep resistance. Meanwhile, the secondary carbides reduce the mechanical properties because they promote intergranular cracking. The control of the type and location of carbides can be applied to control the final mechanical properties after heat treatment.

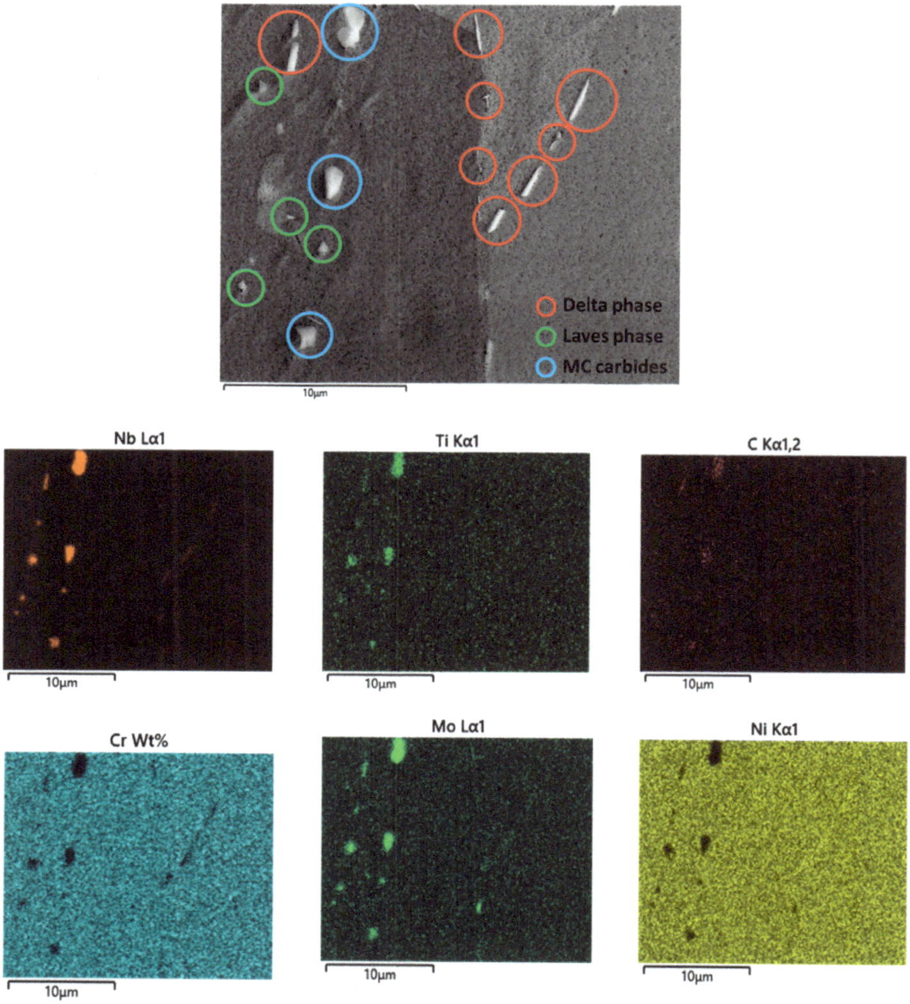

Figure 5. SEM and EDS analysis of the microstructure of IN718. The MC precipitates and Laves, and δ phases were identified through chemical composition, morphology, and distribution.

High cooling rates reduce Nb segregation and the extent of Laves formation. A distinct δ phase is observed at the grain boundaries, characterized by a needle-like morphology with lengths of approximately 1 μm. These findings align consistently with the δ phase prominently detected by DSC measures and the Scheil–Gulliver model simulation at temperatures below 1173 °C.

Figure 6 reveals the presence of γ'', γ', and δ phases despite these not being detected by DSC measures. Laves phases constitute approximately 1 to 2% of the microstructural composition, while NbC carbides contribute 8%. The presence of reinforcing phases γ'', MC carbides, and Laves phases was confirmed using EBSD, SEM, and EDS in interdendritic positions, while the δ phase was found to be located at grain boundaries. This information is consistent with that reported by Hasani et al. [27], who state that the fact δ phase exhibits preferential nucleation at grain boundaries can be attributed to the higher concentration of Nb in these regions. Additionally, δ phase growth predominantly occurs along the grain boundaries, manifesting as needle-like structures, and is observed in a limited number of grain interiors at lower temperatures.

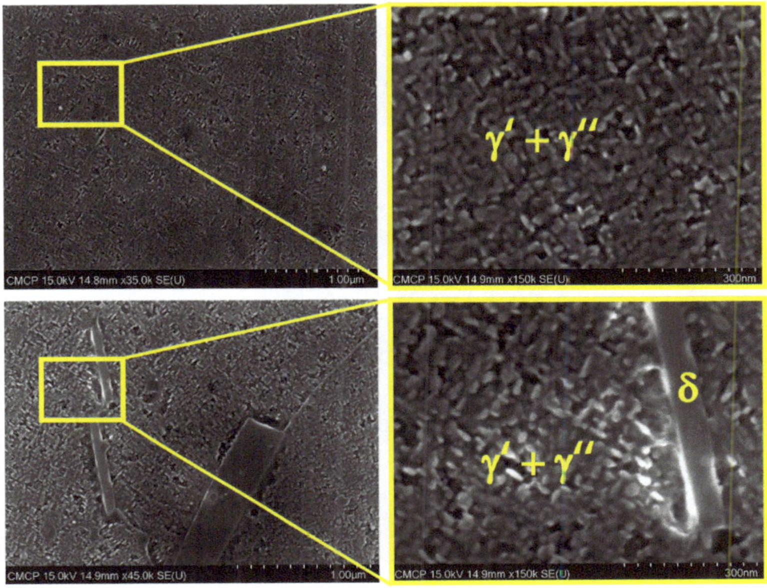

Figure 6. High-magnification SEM images provide a detailed view of the microstructure of IN718. Notably, reinforcing phases are discerned at the nanometer scale, contrasting with the larger-scale δ phase.

EBSD analysis (Figure 7) was conducted to identify Ni_3Nb, Laves, and NbC carbides within the microstructure. The γ'', γ', and the matrix were collectively referred to as Ni_3Nb due to their similar FCC structure and lattice parameters. The coherence between these phases limits the EBSD technique's ability to determine their distribution and crystal orientation accurately [28]. A comparison with SEM morphological shapes from previous studies was carried out to confirm the presence of Laves and NbC phases [14]. For identification, the lattice parameters in Å were used: Ni_3Nb (a = 3.62, b = 3.62, c = 7.41), Laves (a = 5.01, b = 5.01, c = 8.06), and NbC (a = 4.40, b = 4.40, c = 4.40). The IPF images revealed a strong orientation of Ni_3Nb on the (010) plane within the grains. Moreover, the grain boundaries exhibited a higher concentration of NbC and Laves phases with distinct orientations. SEM imaging facilitated the observation of grain boundaries and anisotropic grains. The phase distribution analysis showed that NbC carbides were more prevalent within the grains compared to Laves phases.

The presence of the δ phase at the grain boundary negatively impacted the deformation capacity, resulting in stress/strain concentration around the grain boundary region, as discussed by Rielli et al. [23].

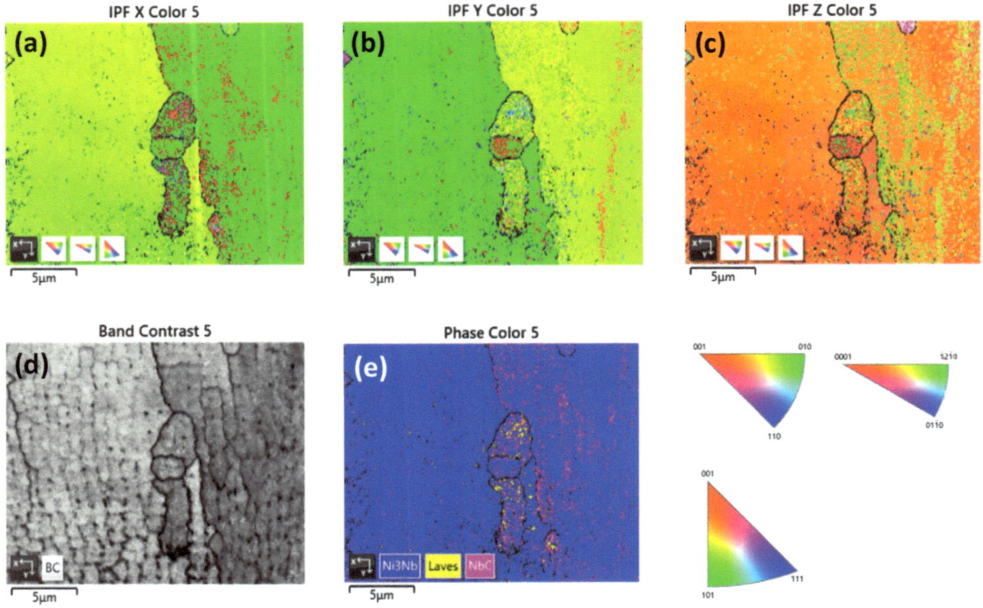

Figure 7. EBSD inverse pole figure (IPF) in three axes (**a**) x, (**b**) y, (**c**) z. (**d**) SEM image, and (**e**) phase distribution of γ'', Laves, and NbC carbides obtained at 20 Kv and tilt 70°.

3.3. TTT Diagram

Several studies, including [29,30], have extensively investigated the TTT (time–temperature transformation) phase diagrams of IN718 to understand the precipitation kinetics of δ and γ'' phases. A TTT diagram determines the dominance of either the driving force or atomic mobility in the formation of phases and their volume fraction [31].

Accurately predicting the phase precipitation is essential for controlling microstructure and mechanical properties, regardless of the manufacturing process. However, the segregation of Nb is significantly influenced by the diffusion of alloys in both interdendritic and dendritic regions, as reported here and by other authors [32,33]. Therefore, it is imperative to consider the thermal history, the initial microstructure, and the chemical composition to construct a reliable TTT diagram that accurately predicts the precipitation of phases.

Figure 8 compares the precipitation of the stable δ phase and the metastable γ'' phase based on temperature and exposure time. The curves show no apparent difference in the shape of the "nose" between the spherical (continuous line) and needle-like (dotted line) morphologies. Moreover, it provides a comparison between the TTT diagram obtained in this work for AM and the TTT diagram for conventional manufacturing published by Thompson et al. [34]. The difference between the two TTT diagrams results mainly from the varying chemical compositions and the methods of manufacture. These factors influence the thermal gradient and dendritic growth in the microstructure, thus modifying the position of the nose and shape of the profile by shifting it to the left of the diagram, which indicates faster transformations due to the rapid cooling that occurs during solidification in additive manufacturing.

The generation of TTT diagrams is affected by various solidification conditions, such as the cooling rate, the dendritic structure, the precipitation kinetics in dendritic and inter-dendritic zones, the availability, and local variations of chemical elements for precipitation, which play significant roles in shaping and positioning the nose of the TTT diagram [32,33]. The TTT phase diagram proposed in this study is a valuable reference tool for comprehending and predicting the solidification process, offering critical insights for controlling and enhancing microstructural properties in additive-manufactured IN718.

By accounting for a range of thermal and temporal parameters, this diagram facilitates informed decision-making throughout the manufacturing process, ensuring precise control over desired phase transformations and ultimately enhancing the overall performance of IN718 components produced via additive manufacturing.

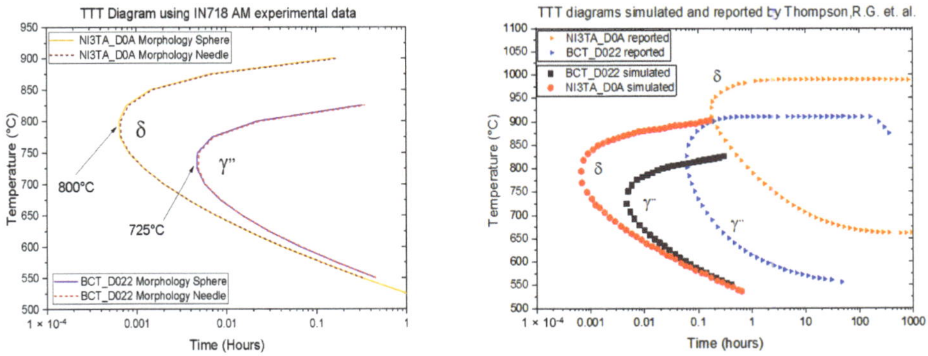

Figure 8. (**Left**) δ and γ'' TTT diagram of IN718 obtained via AM according to morphology shape (sphere and needle) when nucleating in bulk. (**Right**) Comparison between our simulated TTT diagram of IN718 and others referenced [34].

4. Conclusions

In this study, we conducted both experimental investigations and simulations to gain deeper insights into the thermodynamic and microstructural behavior of the IN718 alloy produced via additive manufacturing. Our findings have led to several conclusions:

1. The study successfully predicted the presence of stable and metastable phases and precipitates for a specific chemical composition of the alloy. However, the complex PBF manufacturing process induced intricate phase formations, including Laves phases, which competed with stable and metastable phases such as δ and γ'', ultimately influencing the chemical composition of the matrix.
2. Using DSC analysis and correlation with the Scheil–Gulliver solidification model, we highlighted the predominant formation of MC carbides, a result that was further supported by experimental findings. The presence of Laves phases had significant implications for the availability of chemical elements. Although quantifying the exact proportions of γ'' and δ phases in a specific range proved challenging, the correlations provided valuable insights.
3. Using SEM/EDX/EBSD, it was possible to successfully morphologically identify phases and precipitates based on their distinctive shapes, chemical compositions, and spatial distribution within the microstructure. The complexity of the AM production process significantly influenced the microstructure's morphological and crystalline features, emphasizing the need for a thorough understanding of these conditions to assess the mechanical properties accurately. The confirmation of Laves, MC carbides, and δ phases through various analytical techniques underscored the complexity of the microstructure.
4. Furthermore, we showed that a TTT diagram accurately predicts the solidification process of the δ and γ'' for IN718 superalloy under AM conditions. The comparison between the precipitation of the stable δ phase and the metastable γ'' highlighted the limitations of relying solely on TTT diagrams to depict the complete kinetic behavior of AM-produced samples. Local variations in chemical composition and other factors influenced thermal gradients and dendritic growth in the microstructure, calling for further research to comprehensively understand additive manufacturing and accurately evaluate its mechanical properties.

5. The tailored TTT diagram provides critical information about the phase transformations and microstructural evolution that occur during the heat treatment of components produced by additive manufacturing. With this knowledge, engineers can design heat treatment cycles specifically suited for the AM-produced components. By adjusting the temperature, holding times, and cooling rates based on the TTT diagram, they can optimize the microstructure to achieve the desired material properties, such as strength, hardness, and ductility.

Our comprehensive approach, combining experiments and simulations, has significantly advanced the understanding of the IN718 alloy's behavior in additive manufacturing settings. The knowledge gained from this study can contribute to controlling the microstructural properties of AM materials for various engineering applications.

Author Contributions: Conceptualization, J.A.V.-V. and E.M.-F.; methodology, J.A.V.-V.; software, J.C.F.-C.; validation, J.A.V.-V., J.M.S.-L. and E.M.-F.; formal analysis, J.C.F.-C.; investigation, J.C.F.-C.; data curation, J.C.F.-C.; writing—original draft preparation, J.C.F.-C. and J.M.S.-L.; writing—review and editing, J.A.V.-V. and E.M.-F.; visualization, J.C.F.-C.; supervision, J.A.V.-V.; project administration, C.E.C.-G.; funding acquisition, J.A.V.-V.; project administration, C.E.C.-G. All authors have read and agreed to the published version of the manuscript.

Funding: This research and the APC were funded by CONAHCYT, Mexico. Grant number 000000000319522.

Institutional Review Board Statement: Not applicable.

Informed Consent Statement: Not applicable.

Data Availability Statement: Data is contained within the article.

Acknowledgments: J.C. Franco-Correa thanks Mazen Samara and Dmytro Kevorkov for their assistance during the experimental work, École de Technologie Supérieure (ÉTS) for the equipment support in preparing metallographic samples. Concordia University and TMG research group under the direction of Mamoun Medraj for the technical and financial support during the academic stay. J.C. Franco-Correa also thanks CONAHCYT, Mexico, for the Ph.D. scholarship support and CIDESI for accessing and using laboratories for experimental development. J.A. Villada thanks CONACYT for the support through the "Researchers for Mexico" program.

Conflicts of Interest: The authors declare no conflict of interest. The funders had no role in the design of the study, in the collection, analyses, or interpretation of data, in the writing of the manuscript, or in the decision to publish the results.

References

1. Tan, C.; Weng, F.; Sui, S.; Chew, Y.; Bi, G. Progress and perspectives in laser additive manufacturing of critical aeroengine materials. *Int. J. Mach. Tools Manuf.* **2021**, *170*, 103804. [CrossRef]
2. Ghanavati, R.; Naffakh-Moosavy, H. Additive manufacturing of functionally graded metallic materials: A review of experimental and numerical studies. *J. Mater. Res. Technol.* **2021**, *13*, 1628–1664. [CrossRef]
3. Chen, S.G.; Gao, H.J.; Wu, Q.; Gao, Z.H.; Zhou, X. Review on residual stresses in metal additive manufacturing: Formation mechanisms, parameter dependencies, prediction and control approaches. *J. Mater. Res. Technol.* **2022**, *17*, 2950–2974. [CrossRef]
4. Cooke, S.; Ahmadi, K.; Willerth, S.; Herring, R. Metal additive manufacturing: Technology, metallurgy, and modeling. *J. Manuf. Process.* **2020**, *57*, 978–1003. [CrossRef]
5. Lindwall, G.; Campbell, C.E.; Lass, E.A.; Zhang, F.; Stoudt, M.R.; Allen, A.J.; Levine, L.E. Simulation of TTT curves for additively manufactured Inconel 625. *Metall. Mater. Trans. A Phys.* **2019**, *50*, 457–467. [CrossRef]
6. Paulonis, D.F.; Schirra, J.J. Alloy 718 at Pratt & Whitney-Historical Perspective and Future Challenges. In *Superalloys 718, 625, 706 and Various Derivatives*; TMS (The Minerals, Metals & Materials Society): Warrendale, PA, USA, 2001; pp. 13–23. [CrossRef]
7. *AMS 5662*; Nickel Alloy, Corrosion and Heat-Resistant, Bars, Forgings, and Rings 52.5Ni-19Cr-3.0Mo-5.1Cb (Nb)-0.90Ti-0.50Al-18Fe Consumable Electrode or Vacuum Induction Melted 1775 °F (968 °C) Solution Heat Treated, Precipitation Hardenable—UNS N07718. SAE International: Warrendale, PA, USA, 2016. [CrossRef]
8. *AMS5383F*; Nickel Alloy, Corrosion and Heat-Resistant, Investment Castings 52.5Ni-19Cr-3.0Mo-5.1Cb(Nb)-0.90Ti-0.60Al-18Fe Vacuum Melted Homogenization and Solution Heat Treated. SAE International: Warrendale, PA, USA, 2018. [CrossRef]
9. Trosch, T.; Strößner, J.; Völkl, R.; Glatzel, U. Microstructure and Mechanical Properties of Selective Laser Melted Inconel 718 Compared to Forging and Casting. *Mater. Lett.* **2016**, *164*, 428–431. [CrossRef]

10. Sanchez, S.; Smith, P.; Xu, Z.; Gaspard, G.; Hyde, C.J.; Wits, W.W.; Ashcroft, I.A.; Chen, H.; Clare, A.T. Powder Bed Fusion of nickel-based superalloys: A review. *Int. J. Mach. Tools Manuf.* **2021**, *165*, 103729. [CrossRef]
11. Gallmeyer, T.G.; Moorthy, S.; Kappes, B.B.; Mills, M.J.; Amin-Ahmadi, B.; Stebner, A.P. Knowledge of process-structure-property relationships to engineer better heat treatments for laser powder bed fusion additive manufactured Inconel 718. *Addit. Manuf.* **2020**, *31*, 100977. [CrossRef]
12. Xi, N.; Ni, Z.; Fang, X.; Zhou, Y.; Tang, K.; Zhang, H.; Huang, K. Role of δ-phase on Mechanical Behaviors of Additive Manufactured Inconel 718: Detailed Microstructure Analysis and Crystal Plasticity Modelling. *Int. J. Plast.* **2023**, *168*, 103708. [CrossRef]
13. Franco-Correa, J.C.; Martínez-Franco, E.; Alvarado-Orozco, J.M.; Cáceres-Díaz, L.A.; Espinosa-Arbelaez, D.G.; Villada, J.A. Effect of Conventional Heat Treatments on the Microstructure and Microhardness of IN718 Obtained by Wrought and Additive Manufacturing. *J. Mater. Eng. Perform.* **2021**, *30*, 7035–7045. [CrossRef]
14. De Jaeger, J.; Solas, D.; Baudin, T.; Fandeur, O.; Schmitt, J.H.; Rey, C. Inconel 718 single and multipass modelling of hot forging. In Proceedings of the Superalloys 2012: 12th International Symposium on Superalloys, Champion, PA, USA, 9 September 2012; John Wiley&Sons Inc.: Hoboken, NJ, USA, 2012; pp. 663–672. Available online: https://hal.science/hal-00771436 (accessed on 1 August 2023).
15. Han, W.; Wan, M.; Zhao, R.; Kang, H.; Rao, Y. Microstructural evolution and mechanical properties of brazed IN718 ultrathin-walled capillary structure using different particulate reinforced filler alloy. *Chin. J. Aero.* **2022**, *35*, 550–564. [CrossRef]
16. *ASTM D3418*; 12 Standard Test Method for Transition Temperatures and Enthalpies of Fusion and Crystallization of Polymers by Differential Scanning Calorimetry. ASTM International: West Conshohocken, PA, USA, 2021.
17. *ASTM E407-07(2015)e1*; Standard Practice for Microetching Metals and Alloys. ASTM International: West Conshohocken, PA, USA, 2015. [CrossRef]
18. Rayner, A.J.; Corbin, S.F. Grain growth activation during supersolidus liquid phase sintering in a metal injection molded nickel-base superalloy. *Mater. Today Commun.* **2021**, *26*, 101769. [CrossRef]
19. Shafiee, A.; Moon, J.; Kim, H.S.; Jahazi, M.; Nili-Ahmadabadi, M. Precipitation behaviour and mechanical properties of a new wrought high entropy superalloy. *Mater. Sci. Eng. A* **2019**, *749*, 271–280. [CrossRef]
20. Shi, X.; Duan, S.; Yang, W.; Guo, H.; Guo, J. Solidification and segregation behaviors of superalloy IN718 at a slow cooling rate. *Materials* **2018**, *11*, 2398. [CrossRef]
21. Miao, Z.J.; Shan, A.D.; Wu, Y.B.; Jun, L.U.; Ying, H.U.; Liu, J.L.; Song, H.W. Effects of P and B addition on as-cast microstructure and homogenization parameter of Inconel 718 alloy. *Trans. Nonferrous Met. Soc. China* **2012**, *22*, 318–323. [CrossRef]
22. Vyazovkin, S.; Koga, N.; Schick, C. *Handbook of Thermal Analysis and Calorimetry: Recent Advances, Techniques, and Applications*; Elsevier: Amsterdam, The Netherland, 2018.
23. Rielli, V.V.; Piglione, A.; Pham, M.S.; Primig, S. On the detailed morphological and chemical evolution of phases during laser powder bed fusion and common post-processing heat treatments of IN718. *Addit. Manuf.* **2022**, *50*, 102540. [CrossRef]
24. Knorovsky, G.A.; Cieslak, M.J.; Headley, T.J.; Romig, A.D.; Hammetter, W.F. Inconel 718: A solidification diagram. *Metall. Mater. Trans. A Phys.* **1989**, *20*, 2149–2158. [CrossRef]
25. Tang, Y.T.; Panwisawas, C.; Ghoussoub, J.N.; Gong, Y.; Clark, J.W.; Németh, A.A.; McCartney, D.G.; Reed, R.C. Alloys-by-design: Application to new superalloys for additive manufacturing. *Acta Mater.* **2021**, *202*, 417–436. [CrossRef]
26. Zhao, Z.; Li, L.; Yang, W.; Zeng, Y.; Lian, Y.; Yue, Z. A comprehensive study of the anisotropic tensile properties of laser additive manufactured Ni-based superalloy after heat treatment. *Int. J. Plast.* **2022**, *148*, 103147. [CrossRef]
27. Hasani, N.; Dharmendra, C.; Sanjari, M.; Fazeli, F.; Amirkhiz, B.S.; Pirgazi, H.; Janaki-Ram, G.D.; Mohammadi, M. Laser powder bed fused Inconel 718 in stress-relieved and solution heat-treated conditions. *Mater. Charact.* **2021**, *181*, 111499. [CrossRef]
28. Cao, M.; Zhang, D.; Gao, Y.; Chen, R.; Huang, G.; Feng, Z.; Poprawe, R.; Schleifenbaum, J.H.; Ziegler, S. The effect of homogenization temperature on the microstructure and high temperature mechanical performance of SLM-fabricated IN718 alloy. *Mater. Sci. Eng. A* **2021**, *801*, 140427. [CrossRef]
29. Qin, H.; Bi, Z.; Yu, H.; Feng, G.; Zhang, R.; Guo, X.; Chi, H.; Du, J.; Zhang, J. Assessment of the stress-oriented precipitation hardening designed by interior residual stress during aging in IN718 superalloy. *Mater. Sci. Eng. A* **2018**, *728*, 183–195. [CrossRef]
30. Lin, Y.Y.; Schleifer, F.; Fleck, M.; Glatzel, U. On the interaction between γ'' precipitates and dislocation microstructures in Nb containing single crystal nickel-base alloys. *Mater. Charact.* **2020**, *165*, 110389. [CrossRef]
31. Zhang, S.; Wang, L.; Lin, X.; Yang, H.; Li, M.; Lei, L.; Huang, W. Precipitation behavior of δ phase and its effect on stress rupture properties of selective laser-melted Inconel 718 superalloy. *Composites Part B Eng.* **2021**, *224*, 109202. [CrossRef]
32. Jang, J.; Van, D.; Lee, S.H. Precipitation kinetics of secondary phases induced by heat accumulation in the deposit of Inconel 718. *Addit. Manuf.* **2022**, *55*, 102831. [CrossRef]
33. Yu, X.; Lin, X.; Tan, H.; Hu, Y.; Zhang, S.; Liu, F.; Yang, H.; Huang, W. Microstructure and fatigue crack growth behavior of Inconel 718 superalloy manufactured by laser directed energy deposition. *Int. J. Fatigue* **2021**, *143*, 106005. [CrossRef]
34. Thompson, R.G.; Dobbs, J.; Mayo, D. The effect of heat treatment on microfissuring in alloy 718. *Weld J.* **1986**, *65*, 299–304.

materials

Communication

Fabricating Inner Channels in Laser Additive Manufacturing Process via Thin-Plate-Preplacing Method

Junke Jiao [1,*], Shengyuan Sun [1], Zifa Xu [2], Jiale Wang [1], Liyuan Sheng [3,*] and Jicheng Gao [1]

1 School of Mechanical Engineering, Yangzhou University, Yangzhou 225009, China; mz120210866@stu.yzu.edu.cn (S.S.); mz120220915@stu.yzu.edu.cn (J.W.); 006382@yzu.edu.cn (J.G.)
2 Laser Institute, Qilu University of Technology (Shandong Academy of Sciences), Jinan 250353, China; xuzifatrubo@163.com
3 PKU-HKUST ShenZhen-HongKong Institution, Shenzhen 518057, China
* Correspondence: jiaojunke@yzu.edu.cn (J.J.); lysheng@yeah.net (L.S.)

Abstract: This paper presents a hybrid manufacturing process for the preparation of complex cavity structure parts with high surface quality. Firstly, laser precision packaging technology is utilized to accurately connect a thin plate to a substrate with microchannel. Secondly, Direct Metal Laser-Sintering (DMLS) technology is utilized to completely shape the part. The morphology and microstructure of laser encapsulated specimens and DMLS molded parts were investigated. The results show that the thin plate and the substrate can form a good metallurgical bond. The lowest surface roughness of the DMLS molded parts was 1.18 μm. The perpendicularity between the top of the microchannel and the side wall was optimal when the laser power was 240 W. Consequently, the hybrid manufacturing process effectively solves the problems of poor surface quality and powder sticking of closed inner cavities. The method effectively eliminates the defects of adhesive powder in the inner cavity of the DMLS microchannel, improves the finish, and solves the problem that mechanical tools cannot be processed inside the microchannel, which lays the foundation for the research of DMLS high-quality microchannel process.

Keywords: hybrid manufacturing; surface quality; microstructure; roughness

1. Introduction

Laser Powder Bed Fusion of Metals (PBF-LB/M) technology has become one of the most promising advanced technologies due to its advantages of near-net shaping [1–3]. Taking a high-energy laser beam as the energy source, it realizes the rapid and accurate molding of three-dimensional parts through layer-by-layer superposition [4]. It is gradually applied to all kinds of parts preparation. Therefore, researchers are increasingly focusing on the preparation process of various complex shapes and high quality parts. However, the dimensional accuracy, geometric precision, and surface quality of DMLS parts are not as good as conventionally machined parts, which has hindered the widespread use of this new method [5–7]. One of the key issues with this technology is the deleterious surface quality of the produced parts, which usually requires post-processing. To solve this problem, post-processing such as milling, sandblasting, and polishing are usually performed [8–10].

Many scholars have proposed additive and subtractive hybrid manufacturing technology as a solution to address this problem. Gong et al. [11] utilized additive/subtractive hybrid manufacturing technology to investigate the densification level, microstructure, micro-hardness, and residual stress characterization in different zones of the part by manufacturing 316L SS specimens. Liou et al. [12] proposed a multi-axis laser cladding hybrid processing method that combines a five-axis laser cladding system with a mechanical milling machining center to achieve additive manufacturing of arbitrarily complex shapes by rotating the table, which improves the processing efficiency. Liu et al. [13] developed a hybrid selective laser melting (SLM)/CNC milling system and validated it using stainless steel 316L, resulting in nearly full dense parts with improved dimensional accuracy and surface roughness. Li et al. [14] proposed a six-axis robotic arm with additive manufacturing and subtractive-machining heads to achieve the additive and subtractive hybrid manufacturing process. Soshi et al. [15] demonstrated a rapid fabrication of an injection mold with conformal cooling channels using laser deposition and mechanical machines, improving the cooling performance. Yan et al. [16] successfully produced TC4 thin-wall parts with good build strategy through the hybrid manufacturing process. In the study by Liu et al. [17,18], hybrid additive manufacturing was proposed to improve the surface quality of additive manufacturing components, and it was found that laser polishing greatly reduced the surface roughness and improved the mechanical properties.

Yang et al. [19] successfully prepared high-quality 316L parts using a combination of laser directed energy deposition additive and hot milling cut-off technologies. Through the optimization of process parameters, the microstructure of the parts was refined, and the density was high, showing high micro-hardness (246.73 HV) and tensile strength (683.3 MPa). Du et al. [20] emphasized that additive/subtractive composite manufacturing technology is an effective method for producing high-performance and complex aerospace parts. They specifically applied this technology to the production of martensitic aging steel parts with high density, surface quality, shape, and size accuracy. Jeng et al. [21] developed a new process by combining the selective laser cladding process with the traditional milling process for mold manufacturing and repair. Pan et al. [22] proposed the plasma deposition and milling composite manufacturing process, which greatly enhanced processing efficiency and accuracy by removing plasma deposition layers using conventional milling in a layer-by-layer manner. Additionally, Tian et al. [23] successfully designed and implemented a five-axis additive/subtractive composite processing equipment, enabling the additive/subtractive composite manufacturing of three-dimensional metal parts, including high-temperature alloys, high-entropy alloys, and titanium alloys.

The above research mainly follows the approach of combining laser manufacturing with traditional mechanical processing. However, there have been few studies conducted on the hybrid laser manufacturing of complex inner surface parts, particularly for closed cavities with intricate and precise shapes. Laser processing can effectively address issues such as severe tool wear and poor resolution. The stability of the machining process and the overall quality control of complex internal cavity parts, however, are difficult to achieve.

Therefore, we have investigated the impact of DMLS on thin-walled packages and its effects on the surface quality and microstructure of the top region within the inner cavity. In this paper, a hybrid process of laser precision packaging and DMLS is used to prepare microchannel. Firstly, the cross-sectional morphology of the sheet combined with the substrate after laser packaging was analyzed. Secondly, the effects of four sets of DMLS laser power on the morphology and microstructure of the microchannel inner wall were investigated.

2. Materials and Methods

2.1. Materials

316L SS powders (Chengdu Huayin Powder Technology Co., Ltd., Chengdu, China) were manufactured through gas-atomization and used in the DMLS manufacturing parts. The chemical composition of the powders is provided in Table 1. Figure 1 demonstrates the surface morphology and particle size distribution of the 316L SS powders, with the majority of the powders exhibiting spherical shapes. Figure 1 indicates that 80% of the fine powder had a diameter range of 18 μm to 33 μm, with an average particle size of D determined as 26.03 μm. Dry the powder to 80 °C and keep warm for two hours before the experiment to ensure that the powder does not contain moisture, to avoid the influence of DMLS. The substrate is a 316L plate with micro-grooves of 1 mm depth and width on the substrate. A step of 0.1 mm in depth and 0.05 mm in width was reserved on both sides of the microgroove. The pre-placed thin plates were 316L with width and thickness of 1.1 mm and 0.1 mm, respectively.

Table 1. Chemical composition of 316L SS powders (mass fraction %).

Element	Cr	Ni	Mo	Mn	Si	C	Fe
316L SS powder	17.09	10.61	2.38	1.17	0.59	0.013	Bal

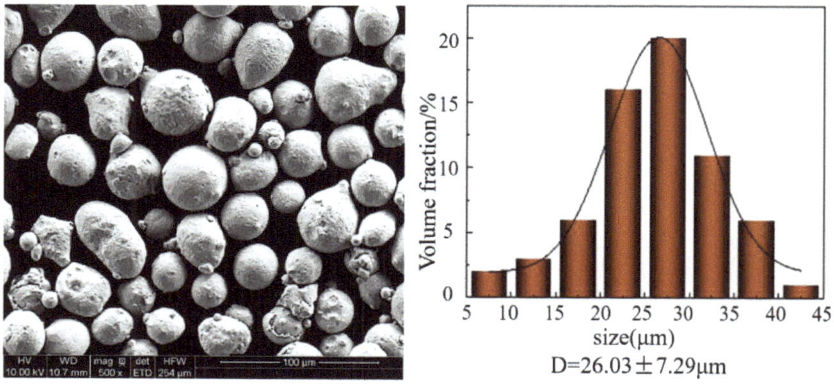

Figure 1. SEM and powder distribution of 316L SS powders.

2.2. *Hybrid Manufacturing Process*

The hybrid manufacturing experiments were carried out using a continuous wave (CW) laser precision packaging system (Maxphotonics, Shenzhen, China) and DMLS system (EOS, Munich, Germany), as shown in Figure 2. The continuous wave laser precision packaging system consists of a continuous wave laser of power P = 1000 W and wavelength λ = 1060 nm, a scanning galvanometer, a motion control system, and an inert gas protection system. DMLS system using EOS-M290 (Beijing Hengshang Technology Co., Ltd., Beijing, China) (Yb-fiber fiber laser peak power P_{pk} = 400 W, spot diameter D = 10–500 µm, maximum scanning speed V_{max} = 7 m/s, layer thickness range 20–100 µm, maximum molding size 250 mm × 250 mm × 325 mm). The experiment was divided into two stages: laser precision packaging and DMLS manufacturing. First, the thin plate was placed on the matrix channel and pressurized to carry out laser precision packaging test. The test parameters are laser power of P = 500 W, a scanning speed of v = 100 mm/s, a defocusing amount of Δf = 10 mm, rotation radius of δ = 1 mm, and a laser diameter of d = 0.15 mm. Then, DMLS manufacturing was performed on the packaged and formed samples, and four groups of samples were prepared by changing the laser power in the test, and the corresponding parameters are shown in Table 2, where P is the laser power, V is the laser scanning speed, Δ is the laser scanning distance, and d is the laser spot diameter. Finally, the wire electrical discharge machining technology is used for cutting and sample preparation. The morphology and microstructure were observed via scanning electron microscope (SEM) FEI Quanta FEG 250(FEI, San Jose, CA, USA) and laser scanning confocal microscope (LSCM) Keyence VK-X200 K (Keyence, Tokyo, Japan).

Table 2. Parameters for DMLS.

No.	P/W	V/mm/s	Δ/mm	d/µm
1	180	920	0.12	30
2	200	920	0.12	30
3	220	920	0.12	30
4	240	920	0.12	30

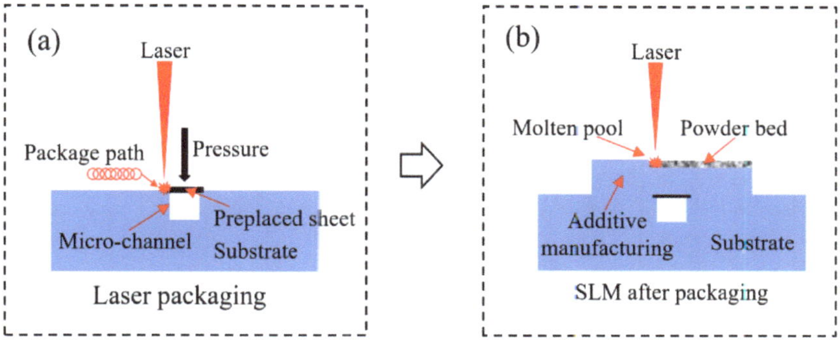

Figure 2. The schematic diagram of hybrid manufacturing process: (**a**) laser precision packaging; (**b**) DMLS after packaging.

3. Results and Discussion

3.1. Surface Topography and Microstructure of the Laser Precision Packaging Specimens

Surface topography and microstructure of the laser precision packaging specimens were examined in this study (Figure 3). We used laser precision packaging to seal the top of the micro-channel on the substrate and to precisely connect the two sides of the micro-channel with a thin plate.

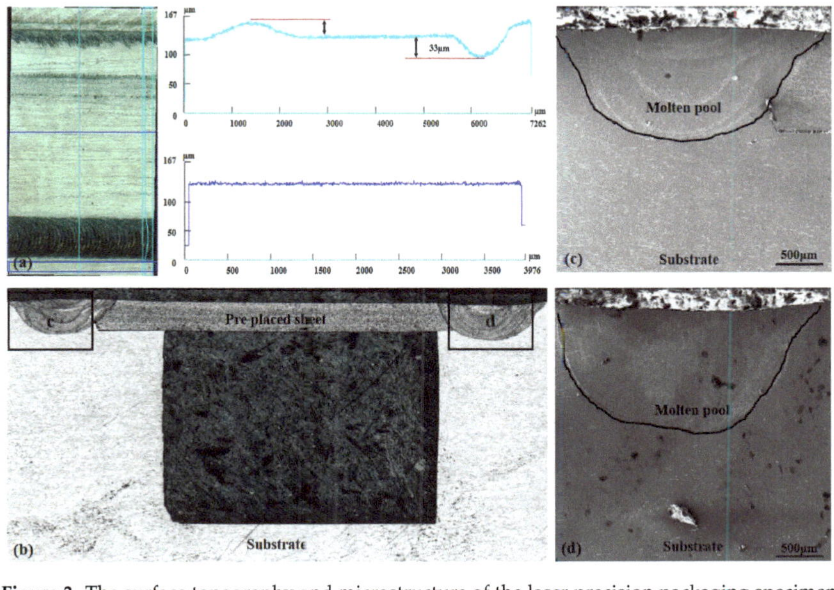

Figure 3. The surface topography and microstructure of the laser precision packaging specimens: (**a**) Thin plate surface morphology; (**b**–**d**) Packaging area micro-morphology.

The analysis results showed that the 0.1mm thin-walled part on the surface of the micro-channel was precisely packaged with better surface quality compared to other areas. However, we observed some irregularities in the form of bumps and depressions in the weld bead, which were attributed to deformation caused by thermal stress in the packaging area. Specifically, we measured a 33 μm deep dent on the right side of the package zone and a 25 μm high bump in the left packaging area experimentally. To assess the microstructure of the packaging area, we utilized scanning electron microscopy (SEM) (Figure 3c,d). The microstructure analysis revealed no notable defects between the prefabricated thin plate and the substrate, with a metallurgical bonding observed between them.

The morphology of the packaging surface, as shown in Figure 3a,b, is relatively flat. The prefabricated thin plate exhibits minimal deformation, with only slight molten pool depressions and protrusions observed in the weld area. This suggests that effective control of thermal stress was achieved during the welding process. Since the DMLS technology can melt the powder together with the substrate and then cool and mold it, the effect of these faint bumpy areas on the subsequent DMLS process is negligible. An analysis of the weld seam (Figure 3c,d) reveals that a metallurgical bond was formed between the prefabricated thin plate and the substrate. Furthermore, almost no pores or crack defects were found in the weld seam, indicating a successful welding outcome.

3.2. *Surface Topography and Microstructure of the DMLS*

The influence of laser power on the top morphology of micro-channels in thin-walled packages produced through DMLS processes was examined after laser precision packaging. Laser powers of 180 W, 200 W, 220 W, and 240 W were employed, as shown in Table 2, to conduct DMLS manufacturing on pre-placed thin plates. Figure 4 presents the outcome of these experiments, illustrating the effect of laser power on the top morphology of micro-channels of 316L SS parts manufactured via hybrid laser precision packaging and DMLS. The experimental findings revealed that the roughness of the top layer of the micro-channel increased as the laser power increased. Specifically, the roughness values (Ra) were 1.18 μm, 1.21 μm, 1.31 μm, and 1.36 μm for power outputs of 180 W, 200 W, 220 W, and 240 W, respectively. The roughness grade reaches approximately Class 7 accuracy, the state in which machining marks can be seen on the surface of machined parts, but not felt by hand. Low surface roughness means a high surface finish, which helps reduce stress concentrations, increase part fatigue strength, and extend service life. However, the change in roughness was not significant, suggesting that the impact of laser power on roughness is relatively small. By observing the top and sidewall profiles of the inner channel shown in Figure 5, it is clear that the top morphology varies under different powers. This variation is a result of the different thermal effects caused by the diverse heating effects of laser inputs at different power levels on the preset thin plate. Notably, at a power of 240 W, the top of the inner channel appears to be the straightest. The performance of the microchannel was affected by its accuracy, and the size accuracy of the microchannel was influenced by the DLMS parameters greatly. To this end, reducing the thermal effect on the microchannel in DMLS process becomes very important. As mentioned above, the best accuracy of the channel can be obtained when the laser power is 240 W, the laser scanning speed is 920 mm/s, the laser scanning distance is 0.12 mm, and the laser spot diameter is 30 μm.

In order to systematically explore the effect of laser power on the overall quality of DMLS manufacturing specimens, LSCM and SEM were used to observe the specimen microstructure. The cross-sectional morphology of the specimens under different laser powers is shown in Figures 6–9. We observed that a good metallurgical bond could be formed between the DMLS zone and the thin plate (Figures 6b, 7b, 8b and 9b). The metallurgical bond allows the molten powder particles to diffuse into the matrix, thus increasing the bond strength between the both. Further magnification of the bonding zone between the DMLS zone and the thin plate reveals that the quality of the boundary of the molten pool after cooling and molding is poor at the power of 180 W and 200 W (Figures 6c and 7c), and pores are present at the boundary of the molten pool at 180 W. Smaller unfused defects are present at the junction of the DMLS zone, the laser packaging zone, and the substrate at 180 W (Figure 6d), which may be caused by poorly bonding the thin plate to the substrate during the laser precision packaging process. It may be caused by a poor fit of the thin plate to the substrate during the laser precision packaging process.

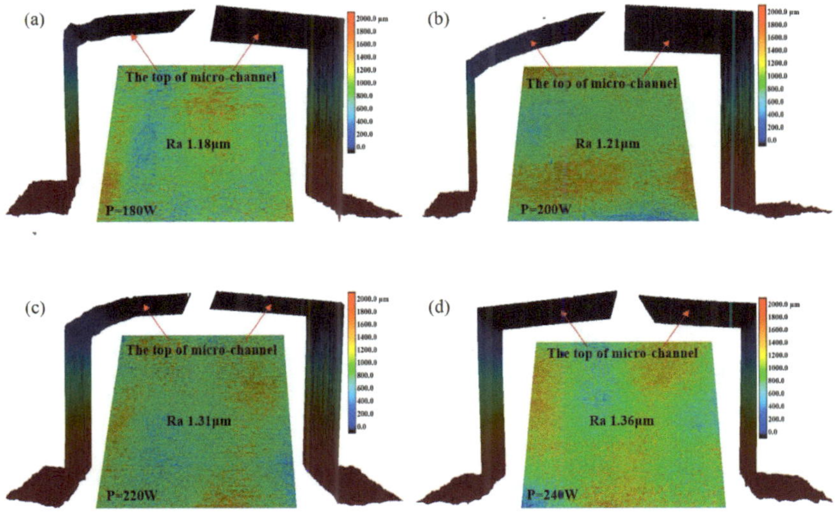

Figure 4. Influence of different laser powers on the top morphology of microchannels in 316L parts: (**a**) 180 W; (**b**) 200 W; (**c**) 220 W; (**d**) 240 W.

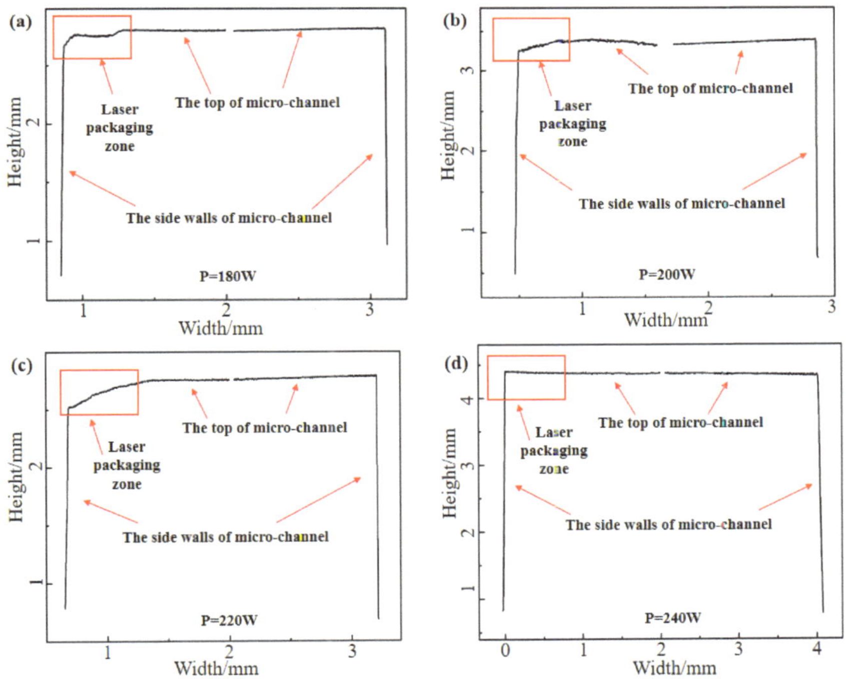

Figure 5. Influence of different laser powers on the microchannel profile of 316L parts: (**a**) 180 W; (**b**) 200 W; (**c**) 220 W; (**d**) 240 W.

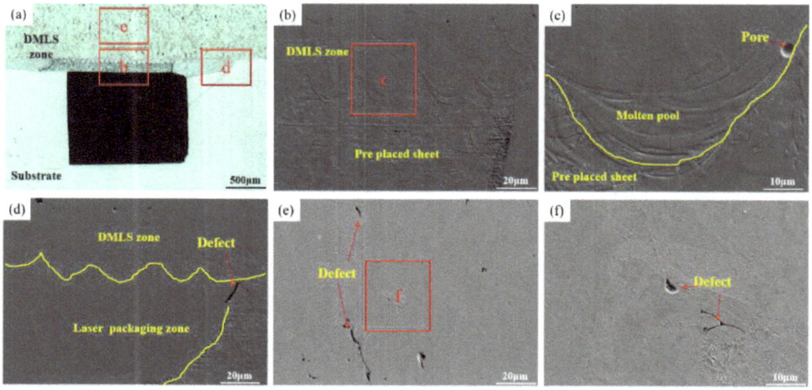

Figure 6. The microstructures when the DMLS laser power is 180 W: (**a**) DMLS areas macro-morphology; (**b**,**c**) Microscopic morphology of the upper layer of the packaging area; (**d**–**f**) DMLS areas with microscopic defects.

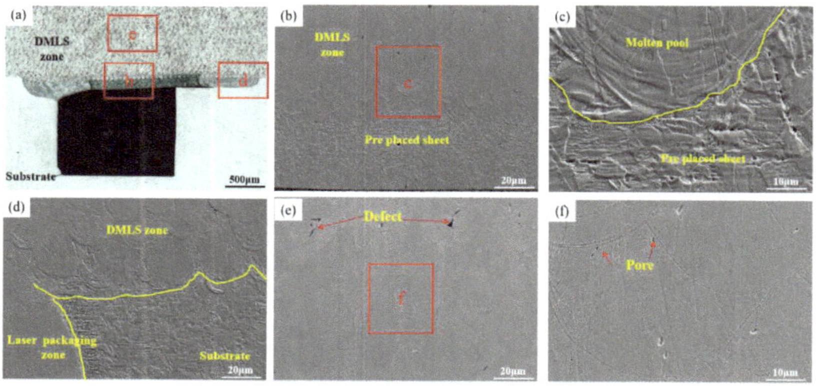

Figure 7. The microstructures when the DMLS laser power is 200 W: (**a**) DMLS areas macro-morphology; (**b**,**c**) Microscopic morphology of the upper layer of the packaging area; (**d**–**f**) DMLS areas with microscopic defects.

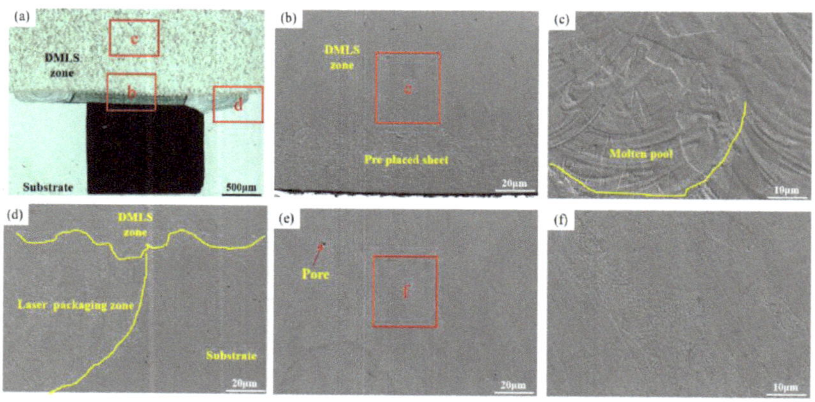

Figure 8. The microstructures when the DMLS laser power is 220 W: (**a**) DMLS areas macro-morphology; (**b**,**c**) Microscopic morphology of the upper layer of the packaging area; (**d**–**f**) DMLS areas with microscopic defects.

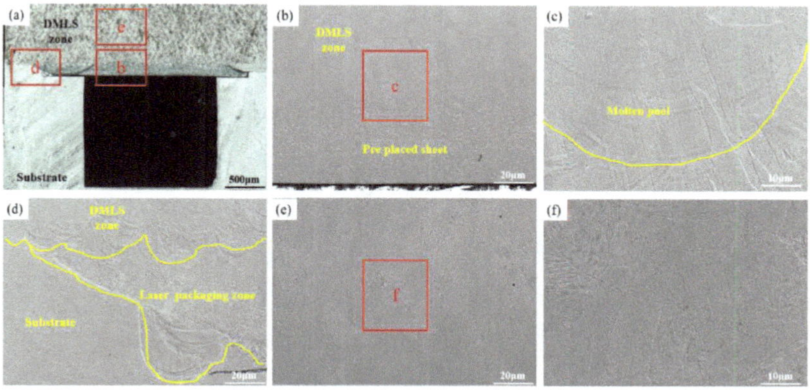

Figure 9. The microstructures when the DMLS laser power is 240 W: (**a**) DMLS areas macromorphology; (**b**,**c**) Microscopic morphology of the upper layer of the packaging area; (**d**–**f**) DMLS areas with microscopic defects.

In addition, the presence of a large number of pores and defects in the DMLS manufacturing zone (Figure 6e) seriously affects the mechanical properties of the parts, which is due to the low heat input. With the increase in laser power, the number of pores and defects in the DMLS zone decreases gradually. When the laser power is 200 W and 220 W, the number of pores and defects in the DMLS layer decreases and becomes smaller, as shown in Figures 7e and 8e. As the laser power was further increased to 240 W, no obvious pores and defects were seen in the sample cross-section (Figure 9e), and its density was the highest among the four samples. The above defects were mainly attributed to the low laser power, which resulted in insufficient heat input and failure of the powder to melt sufficiently, ultimately producing porosity and defects. It is worth noting that the size, number, morphology, and location of the pores have an important effect on the mechanical properties of the components, a higher porosity will shorten the fatigue life of the molded parts, and the pores close to the surface have a greater effect on the fatigue properties of the molded parts than any other location. Therefore, the best quality of DMLS molding was achieved with the laser power of 240 W, scanning speed of 920 mm/s, hatching distance of 0.12 mm, and spot diameter of 30 μm. These pores and defects are mainly caused by the gas in the molten pool escaping too late and the unstable shape of the molten pool. Subsequently, process parameters (scanning speed, spot size, defocusing, etc.) can be optimized to ensure the sealing of the inert gas protected space.

As mentioned above, the DMLS parameters not only influence the accuracy of the microchannel, but also affect the microstructure and the thermal defects of the component, which affect the performance and the mechanical properties of the component. And the best parameter is the laser power of 240 W, scanning speed of 920 mm/s, hatching distance of 0.12 mm, and spot diameter of 30 μm. With this optimized parameter, a microchannel was produced as below. After connecting the thin plate to the two sides of the microchannel via laser precision encapsulation, the microchannel structural member was successfully formed with high quality by using DMLS technology (EOS, Munich, Germany) for manufacturing on the thin plate (Figure 10a). The top surface of the internal cavity does not produce defects such as unmelted particles, the sidewalls are smooth, and the high-quality connection between the top and the side parts can also be realized as shown in Figure 10b. Moreover, the optimization of process parameters minimizes the deformation generated in the packaging area, and the DMLS manufacturing area has no obvious defects and forms a good metallurgical bond with the thin plate and substrate area. The microchannel prepared via this hybrid method has potential applications in the field of microchannels for aero-engine radiators, air-conditioning chillers, water-cooling circuits for large-scale equipment, etc.

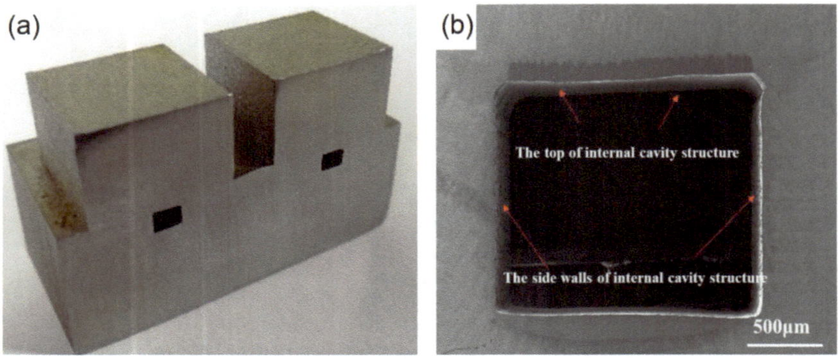

Figure 10. (**a**) Microchannel structural parts fabricated via laser hybrid manufacturing; (**b**) Cross-sectional morphology of microchannel.

4. Conclusions

In this paper, a novel hybrid manufacturing process is proposed to prepare 316L microchannel structural parts with high surface quality. The quality of the parts after laser precision package molding was observed. Then, DMLS manufacturing was carried out on this basis to explore the effects of different process parameters on the top morphology and microstructure of the inner cavity. The main experimental results and analysis of this study can be summarized as follows:

(1) The laser precision packaging process can effectively form a metallurgical bond between the thin plate and the substrate, which improves the bonding strength between the thin plate and the substrate. Under the effect of thermal stress, the left and right sides of the packaging area show a convex mark of 25 μm and a pit of 33 μm, respectively.
(2) By optimizing the DMLS process parameters (P = 180 W, V = 920 mm/s, d = 30 μm, Δ = 0.12mm), the surface roughness of the complex cavity was greatly reduced to Ra 1.18 μm.
(3) When the laser power is 240 W, the top of the internal channel seems to be most perpendicular to the sidewalls. And the DMLS areas have the least defects such as porosity.
(4) The hybrid manufacturing process successfully solved the problems of poor surface quality and powder adhesion in the closed inner cavity, which provides a reference for the research process of the manufactured microchannel.

Author Contributions: Conceptualization, Z.X. and J.J.; methodology, S.S. and L.S.; validation, J.W. and S.S.; investigation, Z.X.; data curation, J.W.; writing—original draft preparation, Z.X. and L.S.; writing—review and editing, J.J. and J.G.; supervision, L.S. and J.G. All authors have read and agreed to the published version of the manuscript.

Funding: The authors gratefully acknowledge the support provided by Ningbo Major Science and Technology Projects (2022Z0013 and 2023Z0031) and Shenzhen Basic Research projects (JCYJ20200109144604020, JCYJ20200109144608205 and JCYJ20210324120001003).

Institutional Review Board Statement: Not applicable.

Informed Consent Statement: Not applicable.

Data Availability Statement: Data sharing is not applicable.

Conflicts of Interest: The authors declare no conflict of interest.

References

1. Martucci, A.; Aversa, A.; Lombardi, M. Ongoing Challenges of Laser-Based Powder Bed Fusion Processing of Al Alloys and Potential Solutions from the Literature—A Review. *Materials* **2023**, *16*, 1084. [CrossRef] [PubMed]
2. Martucci, A.; Aversa, A.; Bondioli, F.; Fino, P.; Lombardi, M. Synergic strategies to improve the PBF-LB\M processability of a cracking-sensitive alloy. *Mater. Des.* **2022**, *224*, 111396. [CrossRef]
3. Mussatto, A. Research progress in multi-material laser-powder bed fusion additive manufacturing: A review of the state-of-the-art techniques for depositing multiple powders with spatial selectivity in a single layer. *Results Eng.* **2022**, *16*, 100769. [CrossRef]
4. Robinson, J.; Arjunan, A.; Baroutaji, A.; Martí, M.; Molina, A.T.; Serrano-Aroca, Á.; Pollard, A. Additive manufacturing of anti-SARS-CoV-2 copper-tungsten-silver alloy. *Rapid Prototyp. J.* **2021**, *27*, 1831–1849. [CrossRef]
5. Li, C.; Liu, D.; Liu, G. Surface characteristics enhancement and morphology evolution of selective-laser-melting (SLM) fabricated stainless steel 316L by laser polishing. *Opt. Laser Technol.* **2023**, *162*, 109246. [CrossRef]
6. Xu, Z.; Ouyang, W.; Jiao, J. Investigation on repairing 30CrMnSiNi2A steel with laser additive and subtractive hybrid remanufacturing technology. *Opt. Lasers Eng.* **2023**, *160*, 107254. [CrossRef]
7. Gora, W.; Tian, Y.; Cabo, A. Enhancing surface finish of additively manufactured titanium and cobalt chrome elements using laser based finishing. *Phys. Procedia* **2016**, *83*, 258–263. [CrossRef]
8. Bobana, J.; Ahmed, A. Improving the surface integrity and mechanical properties of additive manufactured stainless steel components by wire electrical discharge polishing. *J. Mater. Process. Technol.* **2021**, *291*, 117013. [CrossRef]
9. Metelkova, J.; Ordnung, D.; Kinds, Y. Novel strategy for quality improvement of up-facing inclined surfaces of LPBF parts by combining laser-induced shock waves and in situ laser remelting. *J. Mater. Process. Technol.* **2021**, *290*, 116981. [CrossRef]
10. Zhang, S.; Gong, M.; Zeng, X. Residual stress and tensile anisotropy of hybrid wire arc additive-milling subtractive manufacturing. *J. Mater. Process. Technol.* **2021**, *293*, 117077. [CrossRef]
11. Yang, Y.; Gong, Y.; Qu, S. Additive/subtractive hybrid manufacturing of 316L stainless steel powder: Densification, micro-hardness and residual stress. *J. Mech. Sci. Technol.* **2019**, *33*, 5797–5807. [CrossRef]
12. Liou, F.; Slattery, K.; Kinsella, M.; Newkirk, J.; Chou, H.-N.; Landers, R. Applications of a hybrid manufacturing process for fabrication of metallic structures. *Rapid Prototyp. J.* **2007**, *13*, 236–244. [CrossRef]
13. Liu, C.; Yan, D.; Tan, J. Development and Experimental Validation of a Hybrid Selective Laser Melting and CNC Milling System. *Addit. Manuf.* **2020**, *36*, 101550. [CrossRef]
14. Li, L.; Haghighi, A.; Yang, Y. A novel 6-axis hybrid additive-subtractive manufacturing process: Design and case studies. *J. Manuf. Process.* **2018**, *33*, 150–160. [CrossRef]
15. Soshi, M.; Ring, J.; Young, C. Innovative grid molding and cooling using an additive and subtractive hybrid CNC machine tool. *CIRP Ann.* **2017**, *66*, 401–404. [CrossRef]
16. Lei, Y.; Cui, W.; Newkirk, J. Build Strategy Investigation of Ti-6Al-4V Produced Via a Hybrid Manufacturing Process. *JOM J. Miner. Met. Mater. Soc.* **2018**, *70*, 1706–1713.
17. Liu, Y.; Ouyang, W.; Wu, H. Improving surface quality and superficial microstructure of LDED Inconel 718 superalloy processed by hybrid laser polishing. *J. Mater. Process. Technol.* **2022**, *300*, 117428. [CrossRef]
18. Liu, Y.; Sun, S.; Wang, J. Tribological behaviors of LDED Inconel 718 samples polished with a hybrid laser polishing technique. *J. Mater. Res. Technol.* **2023**, *25*, 633–646. [CrossRef]
19. Yang, Y.; Gong, Y.; Li, C.; Wen, X.; Sun, J. Mechanical performance of 316 L stainless steel by hybrid directed energy deposition and thermal milling process. *J. Mater. Process. Technol.* **2021**, *291*, 117023. [CrossRef]
20. Du, W.; Bai, Q.; Zhang, B. Machining characteristics of 18Ni-300 steel in additive/subtractive hybrid manufacturing. *Int. J. Adv. Manuf. Technol.* **2018**, *95*, 2509–2519. [CrossRef]
21. Jeng, J.; Lin, M. Mold fabrication and modification using hybrid processes of selective laser cladding and milling. *J. Mater. Process. Technol.* **2001**, *110*, 98–103. [CrossRef]
22. Pan, Z.; Zhang, H.; Liu, X. Influence of micro-rolling on the strength and ductility of plasma-arc additively manufactured Ti–6Al–4V alloys. *J. Mater. Res. Technol.* **2022**, *21*, 465–473. [CrossRef]
23. Tian, X.; Wu, L.; Li, D. Roadmap for Additive Manufacturing: Toward Intellectualization and Industrialization. *Chin. J. Mech. Eng. Addit. Manuf. Front.* **2022**, *1*, 1100014. [CrossRef]

materials

Article

Mechanical and Surface Characteristics of Selective Laser Melting-Manufactured Dental Prostheses in Different Processing Stages

Edgar Moraru [1], Alina-Maria Stoica [1,*], Octavian Donțu [1], Sorin Cănănău [1], Nicolae-Alexandru Stoica [1], Victor Constantin [1], Daniela-Doina Cioboată [2] and Liliana-Laura Bădiță-Voicu [2]

1 Faculty of Mechanical Engineering and Mechatronics, National University of Science and Technology Politehnica Bucharest, 313 Splaiul Independentei, 060042 Bucharest, Romania; edgar.moraru@upb.ro (E.M.); grigore.dontu@upb.ro (O.D.); sorin.cananau@upb.ro (S.C.); nicolae.stoica@upb.ro (N.-A.S.); victor.constantin@upb.ro (V.C.)

2 The National Institute of Research and Development in Mechatronics and Measurement Technique, 6-8 Soseaua Pantelimon, 021631 Bucharest, Romania; cioboatadoina@yahoo.com (D.-D.C.); badita_l@yahoo.com (L.-L.B.-V.)

* Correspondence: am.stoica@upb.ro

Abstract: Due to the expansion of the use of powder bed fusion metal additive technologies in the medical field, especially for the realization of dental prostheses, in this paper, the authors propose a comparative experimental study of the mechanical characteristics and the state of their microscale surfaces. The comparison was made from material considerations starting from two dental alloys commonly used to realize dental prostheses: Ni-Cr and Co-Cr, but also technologies for obtaining selective laser melting (SLM) and conventional casting. In addition, to compare the performances with the classical casting technology, for the dental prostheses obtained through SLM, the post-processing stage in which they are in a preliminary finishing and polished state was considered. Therefore, for the determination of important mechanical characteristics and the comparative study of dental prostheses, the indentation test was used, after which the hardness, penetration depths (maximum, permanent, and contact depth), contact stiffness, and contact surface were established, and for the determination of the microtopography of the surfaces, atomic force microscopy (AFM) was used, obtaining the local areal roughness parameters at the miniaturized scale—surface average roughness, root-mean-square roughness (RMS), and peak-to-peak values. Following the research carried out, several interesting conclusions were drawn, and the superiority of the SLM technology over the classic casting method for the production of dental prostheses in terms of some mechanical properties was highlighted. At the same time, the degree of finishing of dental prostheses made by SLM has a significant impact on the mechanical characteristics and especially the local roughness parameters on a miniaturized scale, and if we consider the same degree of finishing, no major differences are observed in the roughness parameters of the surfaces of the prostheses produced by different technologies.

Keywords: dental prostheses; biomaterials; additive manufacturing; selective laser melting; indentation test; mechanical characteristics; microtopography; micro-/nano-roughness measurement; atomic force microscopy

1. Introduction

Metal additive technologies from the powder bed fusion (PBF) family open more and more opportunities for modern industries due to their ability to generate the most complicated metal parts with high precision and density in a relatively short time and enhance the final products in terms of geometrical features and in terms of some physical–mechanical properties. Selective laser depositions from the category of PBF additive technologies consist in applying a thin layer of powder with a fine granulation with a special leveling device, after which this powder bed is processed by means of a powerful laser beam or electron beam, melting spherical granules in the place of its projection [1–3]. The selective laser deposition group of technologies includes Selective Laser Sintering (SLS) [4], Direct Metal Laser Sintering (DMLS) [5], and selective laser melting (SLM) [6], which have a similar operating principle, the difference referring to the way the granules are bound. If in the first two mentioned technologies, powdered particles are processed without passing into the liquid phase (they heat up less, and sintering is the base) in the SLM process, more powerful lasers are used leading to the complete melting of the powder particles [1–3,7–9]. With the help of SLM technology, denser structures with better mechanical and surface characteristics are obtained due to the processing conditions and peculiarities, and the field of applicability is a wider one than in the case of other selective laser deposition technologies; but despite all this, the price of SLM equipment is generally higher than DMLS equipment, especially SLS equipment (which usually works with polymers or other non-metallic powders) [1–3,7–9]. All equipment control is performed with the help of a process computer in which the digital model of the future printing structure is loaded. This digital model is preprocessed in the software systems of the equipment in which sacrifice layers are created, scanning and selective processing trajectories are generated for all layers of the structure, important working parameters and specific processing parameters are set for processing a particular material, etc. [1,2]. Electron beam melting (EBM) is an additive processing method that belongs to the PBF category and has working principles similar to selective laser methods. In the case of EBM technology, electron beams are used instead of laser beams to selectively process powdered layers in a vacuum chamber, and only metallic conductive materials can be processed [10]. An alternative to PBF technologies is the metal additive technology, DED (Directed Energy Deposition) [11,12], in which the material in powder or wire form is delivered, melted (with the help of laser beam, electron beams, or arc plasma), and deposited at the same time on a substrate, somewhat similar to the classic additive technology, FDM (Fused Deposition Modelling) [13].

Some of the most frequently used raw materials for the family of PBF technologies are powders from alloys based on cobalt–chromium [14], nickel–chromium [15], various types of steel [16], aluminum [17], noble metals and some of their alloys [18], tungsten [19], titanium and its alloys [20], and powders from non-metallic materials (used more in SLS technology)—polymers [21], composites [22], or even ceramics [23].

The additive technology used to produce the dental prostheses in this paper is the SLM method, which has spectacular strengths and advantages [1–3,8,9,24–27] and is integrated into many important applications and offers new horizons for improving efficiency and capabilities in various fields [28–30], especially the medical field [31–37] and the dental field [1,2,38,39]. Even if the performance of dental prostheses made by SLM technology is worthy of appreciation, there are obviously some disadvantages; for example, the condition of initially obtained surfaces that are rough and require post-processing operations to obtain the right surface quality for the application in which they are provided [1,2,8,9,27]. The problem related to the poor quality of the initial surfaces obtained by selective laser melting technology is solved and compensated with the help of finishing [40] and mechanical [41] or electrochemical polishing methods [42]. These post-processing operations play a decisive role in the durability and future performance of dental prostheses [1,2]. First, there is a very close relationship between roughness and the mechanical properties of the prosthesis; rougher surfaces negatively affect wear resistance and fatigue characteristics, and it is common knowledge that cracks and other mechanical drawbacks occur on the surfaces of prostheses with higher roughness values, and even micro-level defects can contribute to the decrease in several mechanical features and affect the biofunctionality or may even cause the total failure of the prosthetic component [2,43,44]. Furthermore, there are also biological reasons to have a surface with the lowest possible roughness parameters because rougher surfaces facilitate the adhesion and retention of some inopportune and undesired microbial species that can affect the body with various local or even systemic medical problems [2,43,44]. Therefore, it is clear that the post-processing techniques of finishing and polishing the prosthetic components obtained through SLM technology have a fundamental and determining role in minimizing mechanical and biological threats and in increasing durability and mechanical characteristics and fulfilling the functional and anatomical role of the prosthesis [2].

With the advent and accelerated integration of additive manufacturing methods in the medical and dental fields, the scientific community has shown a constant interest in the research topic related to the properties of structures made by these technologies to evaluate, compare with classical technologies, and find viable solutions with the aim to improve the functionality and applicability of prosthetic components. For example, Øilo et al. [45] and Zhou et al. [46] comparatively researched the performances of the resulting mechanical characteristics for the structures made by SLM from a Co-Cr alloy with those made by classical casting and milling methods, and the results in both cases showed a significant influence of the production methods on the mechanical properties, and in the parts made by additive technologies, higher hardness values were obtained compared to the case of classic technologies. Han et al. [47] share the same opinion and conclude that better results can be obtained for dental prostheses using selective laser melting technology than traditional technologies. So, the superiority of SLM technology compared to the classic technologies for the realization of dental prostheses has been demonstrated, but this is also valued due to the corresponding post-processing finishing technologies that were used and contributed to the improvement of the roughness and the establishment of the respective final mechanical properties. Thus, the problem of the roughness of the parts made by SLM is of interest among researchers. Baciu et al. [48] analyzed the quality of the surfaces resulting from sand-blasting operations and observed how the mechanical and surface parameters improved after these operations, considering that 3D printed structures through SLM technology must be post-processed to be compliant for medical applications. In another work, Shu et al. [49] characterized the surface of dental implants made by selective laser melting both at the micro-level with the help of a 3D profilometer, and also at the nano-level with the help of atomic force microscopy.

Considering the lines discussed previously, the main objective of this paper is the evaluation of the mechanical properties and the zonal roughness parameters at the nano-scale level for the dental prostheses obtained by the PBF selective laser melting additive technology in the function of the material, execution, and post-processing technology. In order to assess some mechanical characteristics of dental prostheses, a standardized indentation test detailed in the next chapter was used, after which indentation hardness at specific applied test methodology, deformation level/penetration depths (maximum displacement/maximum indentation depth, zero displacement/permanent indentation depth, and contact depth), contact stiffness, and contact surface were determined. As a comparison term for the indentation test, dental prostheses made by SLM from two different materials, Co-Cr and Ni-Cr alloys and Co-Cr metal structure obtained by casting from a metal–ceramic dental prosthesis, were used. In turn, the dental prosthesis obtained by SLM was tested in two variants—preliminarily finished and polished condition. Regarding the establishment of parameters related to the quality of surfaces at the nano-scale, but also to establish the microtopography of the surfaces of dental prostheses executed by SLM, an AFM microscope was used following the values determined for S_a (surface average roughness), S_q (RMS roughness), and S_y (peak-to-peak values—distance between the extremities of the irregularities). For this purpose, two dental prostheses made of a Co-Cr alloy obtained through SLM (finished and polished) and a dental prosthesis obtained through casting were used. Following the research carried out in this paper, conclusive and interesting results were obtained that demonstrate the strengths of the structures obtained through SLM in terms of mechanical properties and reflect the way in which post-processing operations influence the mechanical and surface characteristics of the dental prostheses made by selective laser melting. Also, mechanical characteristics values and parameters related to surface quality values were determined and compared from several perspectives: the material of the prosthesis (Co-Cr and Ni-Cr), the production technology (SLM and casting), and the stage of post-processing (finished and polished) for the appreciation of the mechanical characteristics, and the production technology (SLM and casting) and the stage of post-processing (finished and polished) to assess the zonal roughness parameters.

2. Materials and Methods

2.1. Materials and Technologies Used

As discussed in the previous chapter, comparative experimental research was performed from three perspectives: depending on the material, the realization technology, and the post-processing technology. Therefore, a total of 4 denture specimens were used for the mechanical indentation tests (Figure 1) as follows:

- Dental crown made by SLM from a Ni-Cr alloy in a preliminary finished state (Figure 1a).
- The mechanical substructure made of a Co-Cr alloy within a metal–ceramic prosthesis, the post-processing state is in polishing conditions and it is practically ready for functional use (Figure 1b).
- Complex prosthesis consisting of several dental structures made by SLM from a Co-Cr alloy in a preliminary finished state. For the mechanical tests, a unitary specimen was used that was sectioned from the entire prosthesis in order not to affect the compliance of the samples and fix them better on the experimental stand platform (Figure 1c).
- Complex prosthesis consisting of several dental structures made by SLM of a Co-Cr alloy in a polished state. For the same reasons as in the case of the finished Co-Cr prosthesis, a unitary specimen of the dental structure was used, which was sectioned from the entire prosthesis (Figure 1d).

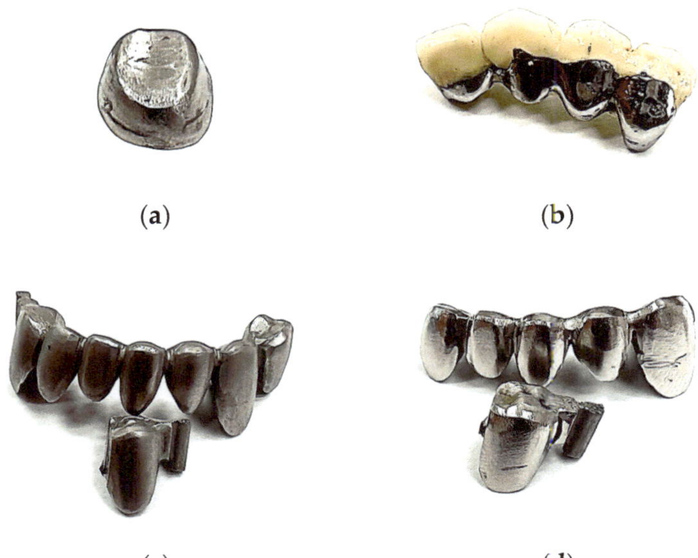

(a) (b) (c) (d)

Figure 1. Dental prostheses used for indentation and surface test: (**a**)—finished SLM manufactured Ni-Cr dental crown; (**b**)—polished Co-Cr dental prosthesis obtained by casting; (**c**)—finished SLM manufactured Co-Cr dental prosthesis; (**d**)—polished SLM manufactured Co-Cr dental prosthesis.

The dental prostheses presented in Figure 1 were executed at an anatomically accurate scale based on the teeth of a real-life patient.

In the case of the tests regarding the microtopography and the quality of the surfaces of the prosthetic components, the same three Co-Cr prostheses were used (the two manufactured by SLM and one by casting); the Ni-Cr dental crown did not intervene in the study because it was considered more evaluation of the prostheses made by SLM regarding the quality of the surfaces depending on the post-processing stage and the comparison with the prostheses made by classical methods. As comparison criteria, we have prostheses made of two different materials, Co-Cr and Ni-Cr, with the same degree of finishing (only for mechanical tests), prostheses made by different technologies, SLM and casting with the same degree of finishing from almost the same material (tests mechanical and surface), prostheses to which different degrees of finishing/post-processing were applied, and finished SLM prosthesis and polished SLM prosthesis (mechanical and surface tests).

The dental prostheses used in this study were made by SLM technology and are based on fine-grained powders of alloys based on Co-Cr (tungsten, molybdenum, and silicon are also contained in the alloy) and alloys based on Ni-Cr (molybdenum, silicon, niobium, and aluminum are also contained in the alloy). After completion, considering the quality of the resulting initial surfaces that left much to be desired, which is a phenomenon specific to powder bed fusion type additive technologies, post-processing operations were gradually applied to them, such as preliminary finishing and mechanical polishing with fine abrasive particles. More details regarding the process of making and post-processing SLM dental prostheses, as well as other particularities, can be found in some previous studies [1,2,50]. Regarding the prosthesis made by classical methods, casting technology was used, and appropriate post-processing operations were used in order to be able to apply the ceramic layer and be ready for biofunctional conditions.

The dental prostheses made with SLM additive technology from materials used in this study were analyzed from the point of view of the elemental composition with the help of scanning electron microscopy (SEM) and energy-dispersive X-ray spectroscopy (EDS) by means of the Thermo Fisher Scientific (Waltham, MA, USA) Phenom ProX analysis system because the structures made do not always fully correspond to the composition specified in the technical data sheet of the raw material [1,2].

Figure 2 represents SEM images (magnification 500×) of one of the investigated points together with the corresponding EDS results for Co-Cr and Ni-Cr dental prostheses realized via SLM [2]. Tables 1 and 2 present the obtained results regarding the average atomic and mass concentrations with one standard deviation (SD) established in six different points on the surface of the cobalt–chromium and five different points on the surface of the nickel–chromium SLM dental prostheses using elemental analysis by EDS spectroscopy [2,51,52].

Table 1. Average atomic and mass concentrations with a standard deviation of a Co-Cr alloy dental prosthesis realized via selective laser melting [2].

	Co	Cr	W	Mo	Si
at%	54.49 ± 1.44	26.15 ± 0.49	9.29 ± 1.54	4.86 ± 0.33	5.21 ± 0.44
wt%	46.65 ± 2.57	19.75 ± 0.92	24.71 ± 3.45	6.76 ± 0.40	2.13 ± 0.20

Table 2. Average atomic and mass concentrations with a standard deviation of a Ni-Cr alloy dental prosthesis realized via selective laser melting [2].

	Ni	Cr	Mo	Si	Nb	Al
at%	63.90 ± 1.76	25.74 ± 0.66	6.30 ± 1.15	2.61 ± 0.71	0.82 ± 0.24	0.63 ± 0.44
wt%	64.02 ± 2.11	22.85 ± 0.63	10.30 ± 1.85	1.25 ± 0.33	1.29 ± 0.38	0.29 ± 0.20

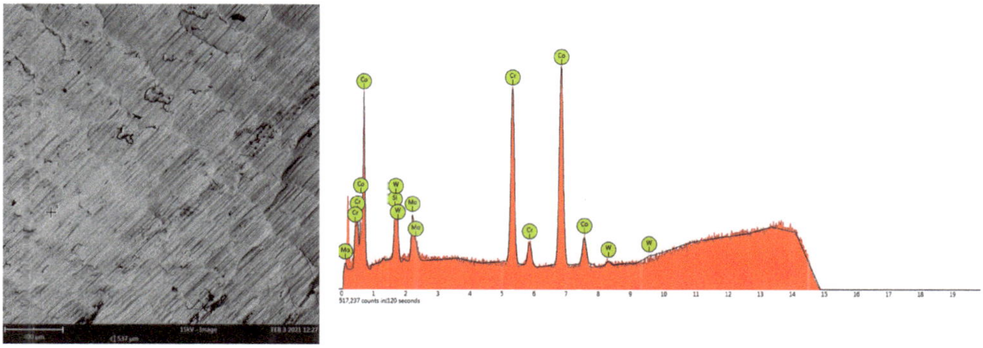

(**a**)

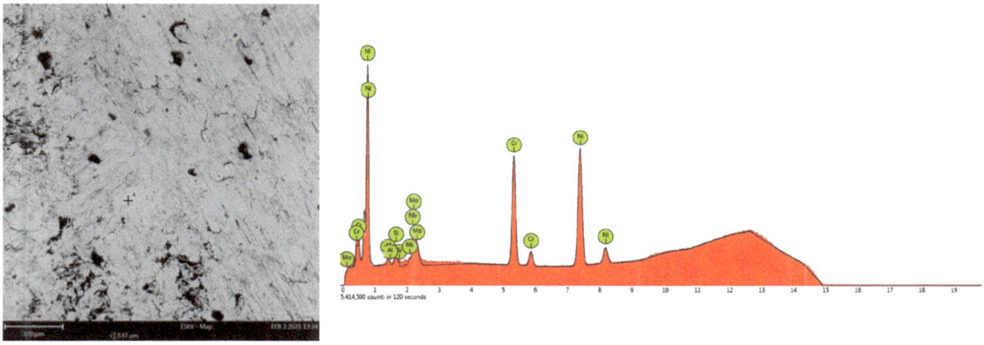

(**b**)

Figure 2. SEM images and EDS results of dental prostheses samples realized via selective laser melting: (**a**)—Co-Cr sample; (**b**)—Ni-Cr sample.

2.2. Indentation Test of Dental Prostheses

Indentation testing, a well-known procedure, was used to determine the mechanical characteristics of the dental prosthesis samples. The tests were performed on a Bruker (Billerica, MA, USA, former CETR (Campbell, CA, USA)) UMT II Multi-Specimen Test System. The UMT II machine was equipped with a DFH-5 2-dimensional force sensor that has a range of 0.5 to 50 N and a resolution of 2.5 mN. In addition, a suspension for the DFH model force sensor was used to maintain the loading force as stable as possible. To measure the displacement, a capacitance sensor with a range of 254 μm and a resolution of 0.01 μm was used. The testing rig setup is presented in Figure 3.

The Indentation tests were performed based on the guidelines of the ISO 14577-1:2015 standard [53]. For these tests, a Rockwell diamond spherical tipped conical indenter that has a 120° angle and a 200 μm radius was used. The selected testing procedure was force-controlled and consisted of three steps, as seen in Figure 4. During the first step, which has 40 s and is called the loading step, the indenter penetrates the material while the applied normal loading force increases constantly for 0 to 25 N. The second step is called the holding stepIse the force is kept constant during a 30 s period. In the third and last step, called the unloading step, the force decreases constantly from 25 to 0 N over a 40 s period. The test force, F_n, the corresponding indentation depth, h, and testing time were recorded during the whole test procedure. All tests were performed at a temperature of 25 °C. To minimize errors, three tests were performed for each sample.

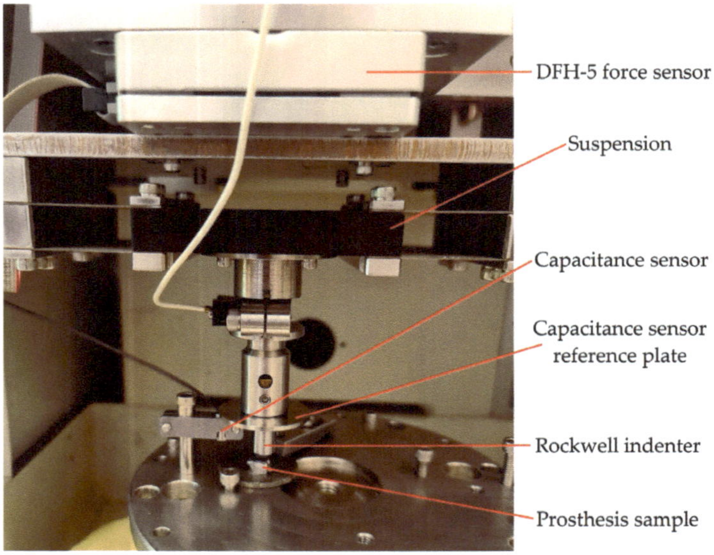

Figure 3. UMT II Multi-Specimen Test System setup.

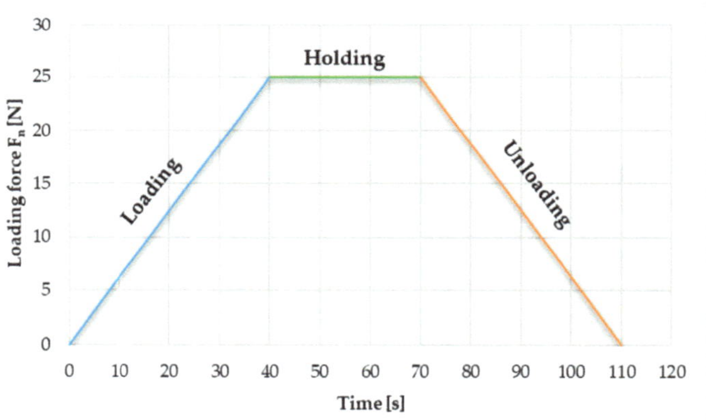

Figure 4. Applied indentation test methodology for dental prostheses.

Based on the normal force and the indentation depth, the post-processing software of the UMT II Test System (Data Viewer 2.16) can automatically determine the indentation hardness (H_{IT}), the maximum displacement, the zero displacement, the contact depth, the contact area, and the contact stiffness ($S = dF_n/dh$).

The indentation hardness, which is a measure of the resistance to permanent penetration or damage, is calculated as [53]:

$$H_{IT} = \frac{F_{max}}{A_p(h_c)}, \quad (1)$$

where F_{max} is the maximum indentation force, $A_p(h_c)$ is the projected (cross-sectional) contact area between the indenter and the test sample, and h_c is the depth of the contact of the indenter with the test sample at F_{max}.

2.3. Surface Characteristics Measurement of Dental Prostheses

Given the influence of denture surface roughness on mechanical characteristics and durability over time, but also biological considerations, issues discussed in Section 1, such as the dental prostheses used in this study, were investigated from the point of view of microtopography, and the main roughness parameters were determined.

For this purpose, we used dental prostheses made by SLM technology from Co-Cr in the finished and polished states to observe how roughness parameters vary depending on the degree of post-processing, but also the Co-Cr metal substructure of the metal–ceramic prosthesis realized through classical casting technology by comparing and analyzing whether the method of obtaining has an impact on the condition of surfaces.

The microtopographic analysis and determination of the main roughness parameters of the dental prostheses took place with the help of an NT-MDT Spectrum Instruments (Limerick, Ireland) NTEGRA Probe NanoLaboratory AFM microscope; the prosthesis on the equipment platform used is presented in Figure 5. This type of microscopy is based on determining the force between a small tip and the area of interest whose roughness parameters are to be established using a cantilever with a sharp tip at the end, and the force acting on the tip after interaction with the area of interest causes the cantilever to bend. By determining the deformation of the cantilever, it is possible to establish the force that occurs at the interaction between the peak and the evaluated area, and by writing down and processing these small deflections of the cantilever, surface topographies can be made by means of the atomic force microscope [2,54,55].

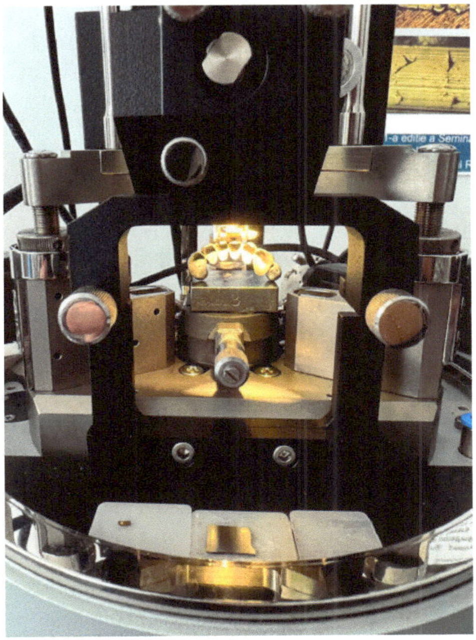

Figure 5. Dental prosthesis on the work platform of an AFM microscope.

The aim of the research is to determine the zonal–local condition of surfaces through microtopography and the nano-roughness of dentures and assess how technology and the post-processing stage influence these aspects at a miniaturized scale. These determinations were based on choosing and scanning 100 × 100 μm areas (from an initial global 800 × 1000 μm area) from different regions on the surface of the 3 prostheses in the study, and then the parameters of interest were evaluated and processed by the AFM equipment software (Nova software). The parameters of interest refer to the areal parameters of roughness: average roughness (S_a), the equivalent arithmetical mean height (R_a parameter of a line) for the surface that expresses the average roughness of the absolute ordinate (vertical) Z axis in a given area (x, y); the root-mean-square parameter of roughness, the *RMS* (S_q) of ordinate (vertical) values in a given area; and the distance between the extremities of irregularities, namely the distance between the highest and the lowest point in a given area, the peak-to-peak value (S_y). Equations (2)–(4) show the relationships with which the previously discussed roughness parameters can be calculated, where A is a given area [56,57].

$$S_a = \frac{1}{A}\iint_A |Z(x,y)|dxdy \tag{2}$$

$$S_q = \sqrt{\frac{1}{A}\iint_A Z^2(x,y)dxdy} \tag{3}$$

$$S_y = max(Z(x,y)) + |min(Z(x,y))| \tag{4}$$

It should be mentioned that the selection of relatively smooth areas at this scale and the avoidance of areas with considerable surface microdefects from the global area initially scanned for all the surfaces of the investigated prosthetic components were taken into account, the final goal was not necessarily to find out the roughness parameters at a global level but at the zonal–local level, and practically to determine the roughness parameters close to the minimum ones in the respective analyzed area and thus see what impact post-processing technologies have at a miniaturized scale by taking into account the problems that may arise due to surface microdefects or biological reasons [2,43,44]. To obtain more conclusive results, three determinations were performed for each sample in different areas of the prostheses.

3. Results

3.1. Results of the Mechanical Characteristics of Dental Prostheses Obtained after an Indentation Test

Based on the indentation tests, we were able to determine the following mechanical properties of the investigated dental prostheses: indentation hardness, H_{IT} 25/40/30/40 (where 25 is the test force, in Newtons, 40 is the application time of test force, in seconds, 30 is the holding time of the test force at maximum test force, in seconds, and 40 is the time taken to remove the test force, in seconds), deformation level after indentation, and three parameters of penetration depth—maximum displacement (maximum indentation depth at maximum applied force), zero displacement (permanent indentation depth, which remained in the structure after the removal of the load), and contact depth (depth of the contact of the diamond indenter with the examined structure at the maximum applied load), and also contact stiffness and contact area. Four samples were tested: a dental prosthesis made by SLM from a Ni-Cr alloy with a finished surface, a dental prosthesis made by SLM from a Co-Cr alloy in polished conditions and one with a finished surface, and a prosthesis made from a Co-Cr metal structure obtained by casting from a metal–ceramic dental prosthesis. The tests were performed three times for each sample, and the results are presented in Table 3, where the average value was calculated for each parameter. Regarding the results obtained for the indentation hardness determined under the conditions described above, an average value of 0.841 GPa was obtained for the dental prosthesis made of Ni-Cr in the finished state made by selective laser melting and a comparable value of 1.636 GPa and 1.683 GPa for the average hardness values were obtained for the polished Co-Cr cast prosthesis and the finished SLM Co-Cr prosthesis, respectively. In the case of the dental prosthesis made by SLM in a polished state, a higher average hardness value was obtained—2.252 GPa. For the resistance to indentation deformability, the maximum deformation reached (maximum displacement), the remaining deformation after releasing the load (zero displacement), and the depth of contact of the indenter with the specimen (contact depth) were determined. Regarding the maximum displacement achieved in the structure, the lowest value results from the polished prosthetic component made by selective laser melting—the average value for the three tests was 14.872 μm, which was the most resistant to deformability among the investigated structures. For the other prosthetic components, we obtained average values for maximum displacement as follows: 35.41 μm for the finished Ni-Cr SLM prosthesis, 20.699 μm for the polished Co-Cr cast prosthesis, and 20.154 μm for the finished SLM Co-Cr prosthesis.

Table 3. Resulting mechanical characteristics from the indentation tests of the dental prostheses.

Property	Material	Ni-Cr	Co-Cr	Co-Cr	Co-Cr
	Method	**SLM**	**Casting**	**SLM**	**SLM**
	Post-Process	**Finished**	**Polished**	**Finished**	**Polished**
Hardness [GPa] H_{IT} 25/40/30/40	1	0.866	1.824	1.607	1.892
	2	0.843	1.648	1.7	2.581
	3	0.815	1.437	1.741	2.252
	Average	0.841	1.636	1.683	2.242
Max. displ. [μm]	1	33.789	18.542	20.921	16.433
	2	35.316	20.177	19.875	12.906
	3	37.13	23.378	19.665	15.278
	Average	35.41	20.699	20.154	14.872
Zero displ. [μm]	1	17.759	6.193	7.147	8.279
	2	17.828	7.077	6.682	4.953
	3	18.138	8.077	6.41	5.883
	Average	17.908	7.116	6.746	6.372
Contact depth [μm]	1	24.369	11.177	12.74	10.763
	2	25.079	12.405	12.02	7.829
	3	26.012	14.289	11.728	9.001
	Average	25.153	12.624	12.163	9.198
Contact stiffness [N/μm]	1	1.903	2.536	2.283	3.293
	2	1.752	2.403	2.378	3.678
	3	1.613	2.053	2.353	2.975
	Average	1.756	2.331	2.338	3.315
Contact area [μm^2]	1	28,757.482	13,653.115	15,500.097	13,161.51
	2	29,539.568	15,105.065	14,650.44	9646.015
	3	30,562.315	17,315.117	14,305.369	11,056.454
	Average	29,619.788	15,357.766	14,818.635	11,287.993

In terms of zero displacement, or residual deformation, things are about the same but with much closer differences for the values obtained for the dental prostheses used in the study: 17.908 μm for the finished Ni-Cr SLM prosthesis, 7.116 μm for the polished Co-Cr cast prosthesis, 6.746 μm for the finished SLM Co-Cr prosthesis, and 6.372 μm for the polished SLM Co-Cr prosthesis. Regarding the contact depth results, we obtained the following results: 25.153 μm for the finished Ni-Cr SLM prosthesis, 12.624 μm for the polished Co-Cr cast prosthesis, 12.163 μm for the finished SLM Co-Cr prosthesis, and 9.198 μm for the polished SLM Co-Cr prosthesis.

The lowest average contact stiffness is obtained for the finished Ni-Cr SLM prosthesis—1.756 N/μm, and the highest value for the polished SLM Co-Cr prosthesis—3.315 N/μm. For the polished Co-Cr cast prosthesis and the finished SLM Co-Cr prosthesis, almost equal values are obtained again—2.331 N/μm and 2.338 N/μm, respectively. And finally, for the contact area, the following average values resulted: 26,619.788 μm^2 for the finished Ni-Cr SLM prosthesis, 15,357.766 μm^2 for the polished Co-Cr cast prosthesis, 14,818.635 μm^2 for the finished SLM Co-Cr prosthesis, and 11,287.993 μm^2 for the polished SLM Co-Cr prosthesis.

Also, based on the research carried out, we determined the indentation behavior of the deformation depending on the applied load according to the methodology described previously and presented in Figure 4. Figure 6 shows the indentation depth displacement behavior of the SLM-manufactured dental prostheses in a finished state in all three investigated zones, and Figure 7 presents the indentation depth displacement behavior of the dental prostheses in the polished state in all three investigated zones (the three tests are differentiated according to the style of the curved line—a continuous line for the first test, a broken line for the second test, and a dotted line for the third test).

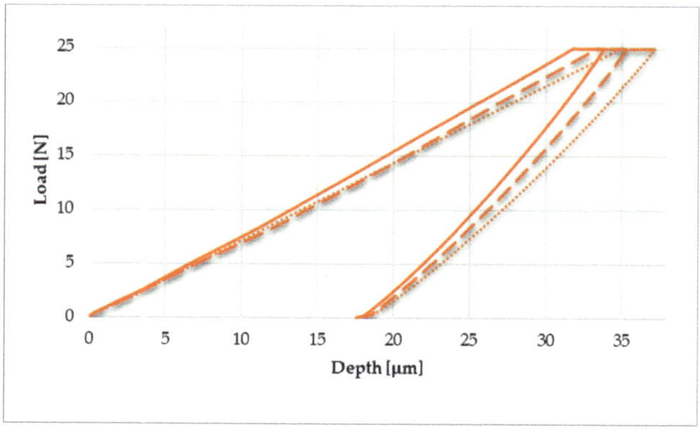

(**a**)

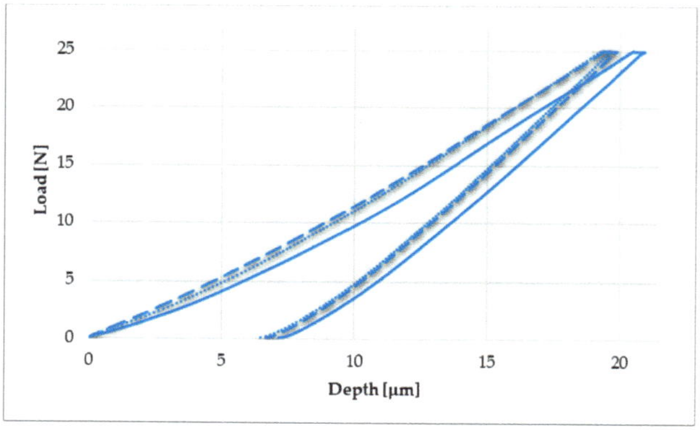

(**b**)

Figure 6. Indentation depth displacement behavior of SLM-manufactured dental prostheses in the finished state—Ni-Cr (**a**) and Co-Cr (**b**) in three different zones (differentiated according to the style of the curved line).

As can be seen in the graphs, there are three distinct areas: when the deformation increases with the force up to 25 N (loading step from research methodology), a smaller area where the deformation increases even though the force remains constant at 25 N (holding step from research methodology), and at the end of this step the maximum displacement value is obtained, and an area where the deformation decreases as the force decreases (unloading step from methodology), finally reaching the residual deformation (zero displacement or permanent indentation depth value). Both in the graphs discussed in this chapter and the following comparative graphs in the Discussions chapter, the same chromatic highlight rule will be kept: brown color for the finished Ni-Cr SLM prosthesis characteristics, purple color for the polished Co-Cr cast prosthesis characteristics, blue color for the finished Co-Cr SLM prosthesis characteristics, and green color for the polished Co-Cr SLM prosthesis characteristics. More details regarding the interpretation of the results and the observations made following the indentation tests and the determination of some mechanical characteristics for dental prosthetic components can be found in the Discussions chapter.

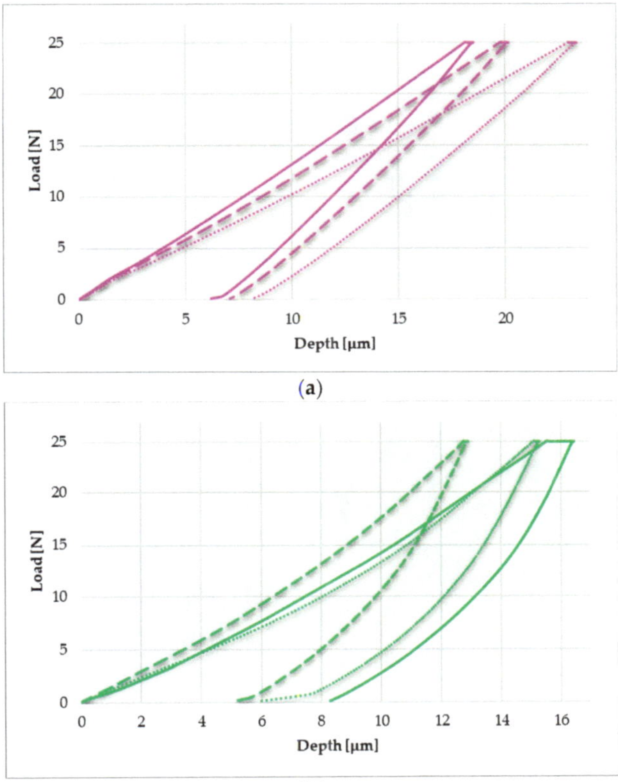

Figure 7. Indentation depth displacement behavior of dental prostheses in the polished state—Co-Cr obtained by casting (**a**) and Co-Cr (**b**) obtained by SLM in three different zones (differentiated according to the style of the curved line).

3.2. *Microtopography and Zonal Nano-Roughness Results of Dental Prostheses Obtained by AFM*

As stated in Section 2.3, the main purpose was to determine the topography and areal roughness parameters at the zonal–local level. Therefore, in the initial phase, an area of 1000 × 800 μm was scanned on the surfaces of the dental prostheses, from which later, through the AFM microscope Nova software, different surfaces of 100 × 100 μm were selected on which the surface evaluations were made. These dimensions of 100 × 100 μm were a criterion applied to all denture surfaces in this study. The equipment program can give us information about the state of the surfaces in 2D format, where the differentiation of higher or lower regions is performed on a color scale, as well as in 3D format, in which the topography of the surfaces can be generated with details about the relief and the distribution of irregularities on the respective.

Figure 8 shows the microtopographies obtained for some areas of 100 × 100 μm (X and Y axes) on the surfaces of the investigated dentures; the surface irregularities were identified on the Z axis. The high peaks are much more pronounced for the SLM dental prostheses in the finished state (Figure 8a) than the polished dental prosthetic components.

There are no major differences between the microtopographic images obtained for the polished SLM prosthesis (Figure 8b) and the cast prosthesis in a polished state (Figure 8c), except that perhaps at the prosthetic structure made by selective laser melting, a smaller difference between the highest and lowest regions on the surface and a more uniform and dense distribution of irregularities on the surface can be observed.

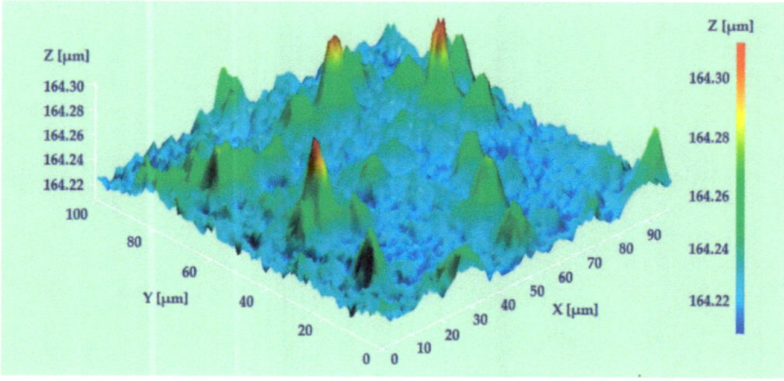

(**a**)

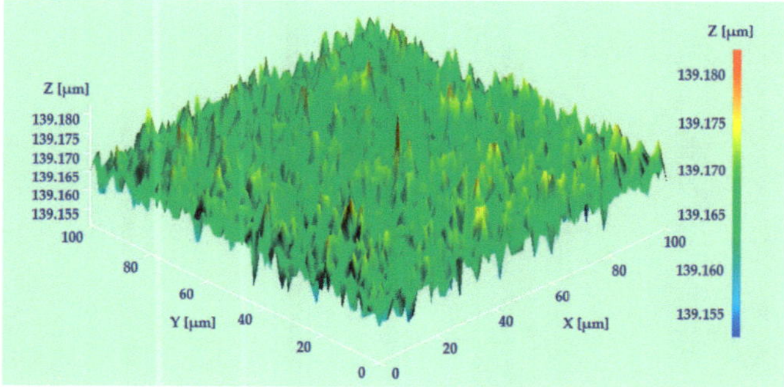

(**b**)

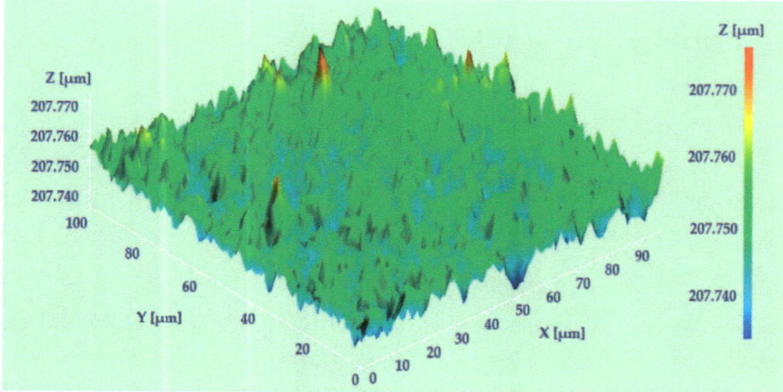

(**c**)

Figure 8. Obtained microtopography of dental prostheses on an AFM microscope: (**a**)—finished SLM manufactured Co-Cr dental prosthesis; (**b**)—polished SLM manufactured Co-Cr dental prosthesis; (**c**)—polished Co-Cr dental prosthesis obtained by casting.

It should be mentioned that the reference point for each investigated area is not zero, it depends on the placement of the prosthesis on the measuring platform of the equipment, and this aspect can be seen in the values on the Z axis from the microtopographic images. This does not affect the values of the established roughness parameters; basically, the difference between the highest and the lowest value on the Z axis is actually the peak-to-peak value (S_y).

Also, following the measurement on the AFM microscope, information can be generated regarding the average profile on the X and Y axes, the histogram of the measurement, and the values of the determined roughness parameters. Table 4 shows the results obtained for the main parameters from zonal nano-roughness determinations on the AFM microscope. In addition to the results illustrated in the table, other parameters can also be determined on the AFM microscope: the maximum and minimum point in the investigated area (the difference between them being the S_y parameter), ten-point height value (S_z), surface skewness (S_{sk}), coefficient of kurtosis (S_{ka}), and other parameters related to the condition of the surfaces [2,54,55]

Table 4. Resulting main parameters from zonal nano-roughness determinations on an AFM microscope.

Parameter	Material	Co-Cr	Co-Cr	Co-Cr
	Method	SLM	SLM	Casting
	Post-Process	Finished	Polished	Polished
Average roughness S_a (nm)	1	6.81	4.53	3.37
	2	9.80	2.78	3.72
	3	6.64	2.95	3.50
	Average	7.75	3.42	3.53
Root-mean-square S_q (nm)	1	9.68	5.62	4.30
	2	15.06	3.46	4.81
	3	9.47	3.67	4.50
	Average	11.40	4.25	4.54
Peak-to-peak S_y (nm)	1	94.03	38.09	38.32
	2	183.23	30.94	45.41
	3	94.03	28.56	42.57
	Average	123.76	32.53	42.10

All the results obtained and illustrated in Table 4 were rounded to two decimal places. Therefore, after calculating the average of the values obtained for the three areas considered in this study, the following results regarding the zonal–local roughness can be highlighted. In the case of the finished SLM prosthesis, there was an average roughness of 7.75 nm, an average RMS value of 11.40 nm, and an average peak-to-peak value of 123.76 nm; in the case of the polished SLM prosthesis, there was an average roughness of 3.42 nm, an average RMS value of 4.25 nm, and an average peak-to-peak value of 32.53 nm; in the case of the polished cast prosthesis, there was an average roughness of 3.53 nm, an average RMS value of 4.54 nm, and an average peak-to-peak value of 42.10 nm. The results obtained regarding the microtopography and nano-roughness of the surfaces of dental prostheses provide interesting conclusions regarding the impact of post-processing operations on the quality of the resulting surfaces, and discussions on this topic are developed in the next chapter.

4. Discussions

4.1. Mechanical Characteristics

A comparative graphical analysis of the load-depth curves for the indentations with the highest displacement (the most affected zone by the indentation) for the four prostheses samples can be seen in Figure 9. These curves indicate a visco-elasto-plastic behavior for all four samples, with an emphasis on the polished SLM Co-Cr sample. Furthermore, it can be clearly seen that the Ni-Cr sample has a much higher deformation than the Co-Cr samples, both in terms of maximum displacement (the penetration depth at the end of the holding stage) and zero displacement (the residual plastic deformation at the end of the unloading stage). Also, the Ni-Cr prosthesis shows the greatest increase in indentation creep deformation (indentation deformation during the holding step). For the Co-Cr samples, it can be also highlighted that plastic deformation is not significantly influenced by post-processing operations, but the elastic deformation (difference between zero and maximum displacement) is. The polished SLM sample has a lower elastic deformation than the finished ones and the casted one, indicating that it is the hardest to deform.

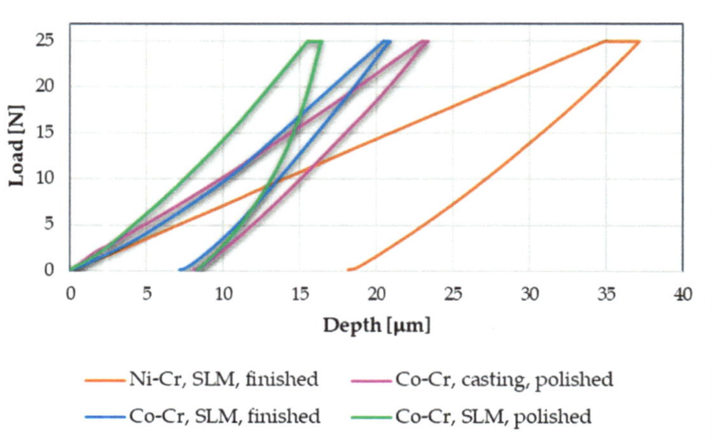

Figure 9. Comparison of indentation load-depth (displacement) curves for the dental prostheses in the most affected zone.

To better analyze the results, a graphical comparison of the average values of the mechanical characteristics obtained from the indentation tests was performed and can be seen in the diagrams presented in Figure 10. As expected, the Co-Cr samples, due to the presence of tungsten in their composition, have better mechanical properties and are deformed less than the Ni-Cr one.

By comparing the average results for the finished and polished SLM prostheses, we can see that the post-processing operation applied to the surface has a significant impact on the indentation behavior of the material. Polishing the surface of the prosthesis leads to a 33.2% increase in hardness and a 41.8% increase in the contact stiffness compared to the finished surface. Regarding the indentation deformations, we can observe that the polishing of the SLM dental prosthesis compared to the preliminarily finishing process has the following effects: a 26.2% decrease in the maximum displacement, a 5.5% decrease in the zero displacement, a 24.4% decrease in the contact depth, and a 23.8% decrease in the contact area.

The method by which the samples are made influences their mechanical properties. This can be seen by comparing the average results of the polished SLM Co-Cr sample and the casted Co-Cr sample. The SLM prosthesis has a 37% higher hardness and a 42.2% higher contact stiffness than that obtained through casting. Furthermore, the polished SLM prosthesis deforms significantly less than the traditionally manufactured one: a 28.1% lower maximum displacement and a 10.5% lower zero displacement, and a 27.1% lower contact depth and a 26.5% contact area than the polished casted one. In terms of mechanical properties and deformation, the polished casted Co-Cr sample has results similar to the finished SLM sample with comparable values for all investigated parameters.

If we consider the same manufacturing technology and the same post-processing step (finishing) by analyzing the two different materials, Ni-Cr and Co-Cr alloys, it can be observed that the Co-Cr alloy has a double indentation hardness and a 33.1% higher contact stiffness compared to the Ni-Cr alloy. Also, in terms of deformation, the Co-Cr sample has a 43.1% lower maximum displacement, a 62.3% lower zero displacement, a 51.65% lower contact depth, and a 50% lower contact area.

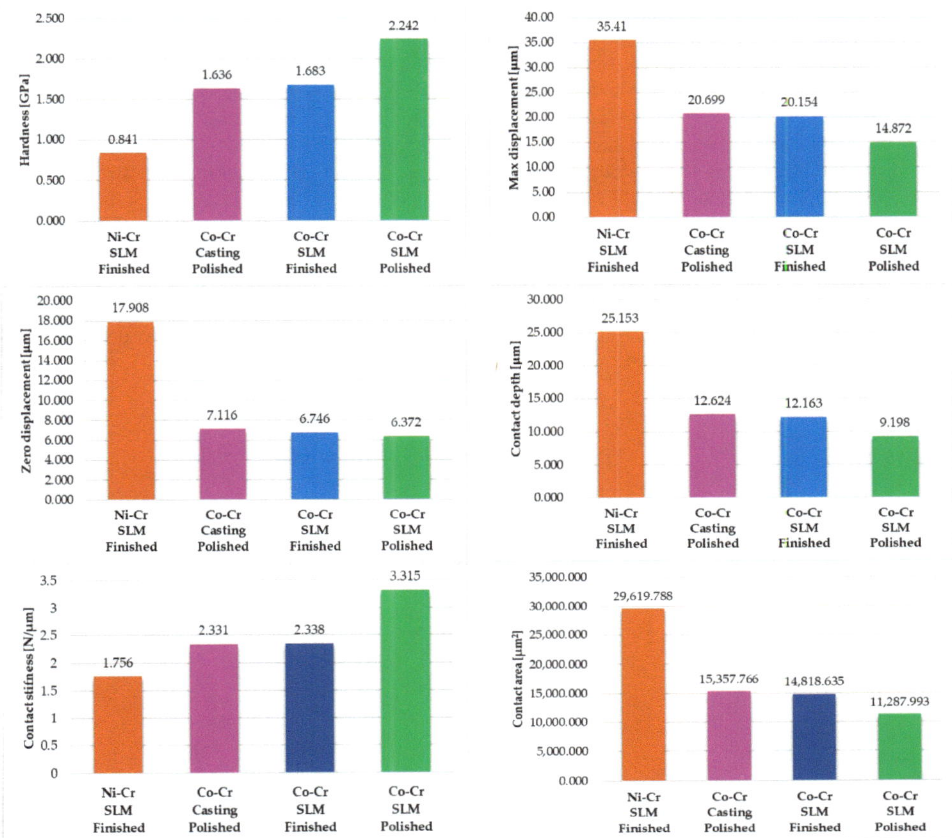

Figure 10. Comparison of the average values of the mechanical characteristics obtained after indentation tests.

After the interpretation of the results of the indentation tests, the superiority of the SLM sample over the casted sample has been demonstrated in terms of mechanical characteristics due to specific particularities of the manufacturing process (the effect of laser layer processing during the selective laser melting technology can also be considered a thermal treatment that enhances some mechanical properties). Moreover, it was shown that the post-processing operations have a major impact on the behavior of the samples during the indentation test and improve their mechanical characteristics.

4.2. Microtopography and Zonal Nano-Roughness

As can be seen in the microtopographic images (Figure 8), the post-processing operations have a significant influence even at this miniaturized scale. Compared to the dental prostheses after the last post-processing operation (polishing) before the application of the ceramic layer, the finished dental prosthesis shows, in the area of 100 μm × 100 μm, major differences between the height of the irregularities and has well-pronounced high peaks. In the case of both polished dental prostheses, one made by additive technology and one by classical casting technology, the result is a surface with a much smoother and denser distribution without significant differences between the extremities of the irregularities. Practically, the polished prostheses have an almost identical distribution of irregularities, perhaps with a slight superiority of the SLM-polished prostheses, observing a slightly smoother and more homogeneous distribution and a slightly smaller difference between the peaks compared to the surface of the casted prosthesis.

Figure 11 compares the results of the surface parameters' average values obtained after zonal nano-roughness determinations on an AFM microscope. Significant differences are observed between the values obtained for the preliminarily finished prosthesis surface and the polished prosthesis surfaces. Regarding the average roughness, it turns out that polishing improves this parameter at this scale more than twice (average values of S_a were 7.75 nm for finished SLM prosthesis, 3.42 nm for polished SLM prosthesis, and 3.53 nm for polished cast prosthesis), and in the case of an RMS, over 2.5 times (average values of RMS were 11.40 nm for finished SLM prosthesis, 4.25 nm for polished SLM prosthesis, and 4.54 nm for polished cast prosthesis). If we are discussing peak-to-peak values, differences are even more accentuated (average values of S_y were 123.76 nm for finished SLM prosthesis, 32.53 nm for polished SLM prosthesis, and 42.10 nm for polished cast prosthesis).

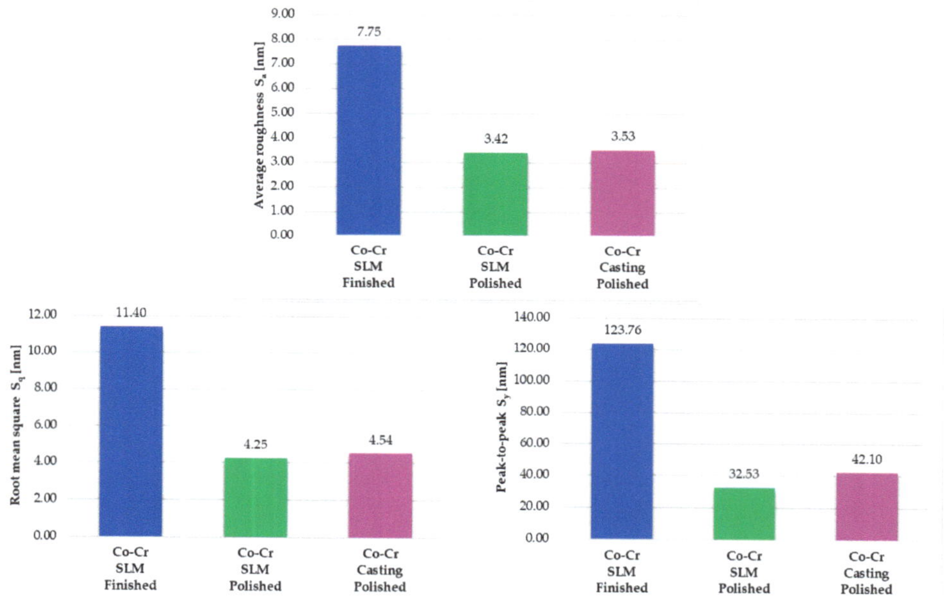

Figure 11. Comparison of surface parameters' average values obtained after zonal nano-roughness determinations on an AFM microscope.

It seems that the method of obtaining does not have a major impact on the state of the resulting surfaces, as long as the same mechanical polishing process is applied. The major differences occur between the surfaces in different stages of post-processing. Relatively comparable values were obtained in terms of average roughness and RMS for the polished prostheses made by SLM and casting. However, there are relatively small differences in the S_y parameter, which refers to the distance between the extremities of the irregularities, where the average value of this parameter was almost 10 nm lower in the case of the dental prosthesis surface obtained by SLM, but it should be noted that in the case of the prosthesis made by casting, the closest results were obtained for the three studied areas (the difference between the maximum and minimum value for roughness parameters) in the three prostheses under study.

Taking into account that the surface roughness of dental prostheses, including at the miniaturized scale, influences their performance from biological considerations or mechanical properties, this study is useful for evaluating the efficiency of the post-processing operations at the micro- or nano-level and for selecting the optimal process or the eventual correction of the one used [2,43,44,48–50].

5. Conclusions

Based on the research carried out in this paper, the following can be concluded:

- Considering the same technology (SLM) and the same degree of finishing, compared to the Ni-Cr prosthesis, the Co-Cr one had a twice higher average indentation hardness and a contact stiffness over 33% higher. The Co-Cr prosthesis was also harder to deform than the Ni-Cr one.
- Considering the same degree of finishing and almost the same material (Co-Cr alloy), the prosthesis realized via SLM has a 37% higher indentation hardness and more than 42% higher contact stiffness than the prosthesis obtained by casting. Also, the indentation deformations are much smaller in the case of the prosthesis made by SLM than in the case of the prosthesis obtained by casting.
- The best mechanical properties and the greatest resistance to deformation after indentation tests were obtained in the case of the polished SLM dental prosthesis. The preliminarily finished SLM dental prosthesis has mechanical properties comparable to the polished one made by casting.
- Both in the case of the mechanical properties and especially in the case of the condition of the surfaces, the major impact that the post-processing operations of dental prostheses have on these characteristics was shown. After polishing the dental prosthesis made by SLM, a significantly higher average indentation hardness was obtained compared to the results obtained for the preliminarily finished SLM dental prosthesis (2.242 GPa compared to 1.683 GPa) with a greater contact stiffness (3.315 N/μm compared to 2.338 N/μm). Also, the polishing of the prosthesis led to an increase in resistance to deformation; for example, an average maximum displacement of 14.872 μm was obtained for the polished sample compared to an average maximum displacement of 20.154 μm for the preliminarily finished sample. The effects of polishing were even more evident for the average values determined for the microscale roughness parameters: in the case of the polished SLM dental prosthesis, the average roughness improved more than twice, the RMS parameter more than 2.5 times, and the distance between the extremities of the irregularities decreased almost 4 times compared to preliminarily finished SLM dental prosthesis.
- The roughness parameters are not significantly influenced by the method of obtaining the dental prostheses after applying the same degree of post-processing. Obviously, the differences appear when moving to a higher degree of finishing.
- Surface roughness plays a decisive role in durability and mechanical properties, but also in avoiding possible problems of biological nature; therefore, the use of appropriate post-processing operations for dental prostheses is very important.

As possible research perspectives, we can mention the study of other materials, technologies, or other post-processing methods for dental prostheses and also use other analysis and characterization methods and equipment.

Author Contributions: Conceptualization, E.M., O.D. and S.C.; methodology, E.M., A.-M.S., N.-A.S. and L.-L.B.-V.; software, E.M., A.-M.S., N.-A.S., V.C. and L.-L.B.-V.; validation, O.D., S.C. and D.-D.C.; formal analysis, E.M., A.-M.S., V.C. and D.-D.C.; investigation, E.M., A.-M.S. and N.-A.S.; writing—original draft preparation, E.M. and N.-A.S.; writing—review and editing, A.-M.S. and V.C.; supervision, O.D., S.C., D.-D.C. and L.-L.B.-V. All authors have read and agreed to the published version of the manuscript.

Funding: This work was funded by the European Social Fund from the Sectorial Operational Programme Human Capital 2014–2020 by the Financial Agreement with the title "Training of PhD students and postdoctoral researchers in order to acquire applied research skills—SMART", contract No. 13530/16.06.2022—SMIS code 153734.

Institutional Review Board Statement: Not applicable.

Informed Consent Statement: Not applicable.

Data Availability Statement: The data presented in this study are available upon request from the corresponding author.

Acknowledgments: This work was funded by the European Social Fund from the Sectorial Operational Programme Human Capital 2014–2020 by the Financial Agreement with the title "Training of PhD students and postdoctoral researchers in order to acquire applied research skills—SMART", contract No. 13530/16.06.2022—SMIS code 153734.

Conflicts of Interest: The authors declare no conflict of interest.

References

1. Moraru, E.; Dontu, O.; Petre, A.; Vaireanu, D.; Constantinescu, F.; Besnea, D. Some technological particularities on the execution of dental prostheses realized by selective laser deposition. *J. Optoelectron. Adv. Mater.* **2018**, *20*, 208–213.
2. Moraru, E. Research on the Realization of Dental Prostheses by Selective Laser Deposition and Other Additive Technologies. Ph.D. Thesis, Politehnica University of Bucharest, Bucharest, Romania, 2021. (In Romanian).
3. Singh, D.D.; Mahender, T.; Reddy, A.R. Powder bed fusion process: A brief review. *Mater. Today Proc.* **2021**, *46*, 350–355. [CrossRef]
4. Lupone, F.; Padovano, E.; Casamento, F.; Badini, C. Process phenomena and material properties in selective laser sintering of polymers: A review. *Materials* **2021**, *15*, 183. [CrossRef]
5. Nandy, J.; Sarangi, H.; Sahoo, S. A review on direct metal laser sintering: Process features and microstructure modeling. *Lasers Manuf. Mater. Process.* **2019**, *6*, 280–316. [CrossRef]
6. Sefene, E.M. State-of-the-art of selective laser melting process: A comprehensive review. *J. Manuf. Syst.* **2022**, *63*, 250–274. [CrossRef]
7. Sharma, S.K.; Singh, A.K.; Mishra, R.K.; Shukla, A.K.; Sharma, C. Processing Techniques, Microstructural and Mechanical Properties of Additive Manufactured 316L Stainless Steel. *J. Inst. Eng. India Ser. D* **2023**, 1–14. [CrossRef]
8. Wenext, DMLS vs SLM—What Is' the Difference? Available online: https://www.wenext.com/blog/3D-Printing/dmls-vs-slm (accessed on 30 July 2023).
9. Xometry, DMLS vs. SLM: Differences and Comparison. Available online: https://www.xometry.com/resources/3d-printing/dmls-vs-slm-3d-printing/ (accessed on 30 July 2023).
10. Körner, C. Additive manufacturing of metallic components by selective electron beam melting—A review. *Int. Mater. Rev.* **2016**, *61*, 361–377. [CrossRef]
11. Svetlizky, D.; Das, M.; Zheng, B.; Vyatskikh, A.L.; Bose, S.; Bandyopadhyay, A.; Schoenung, J.M.; Lavernia, E.J.; Eliaz, N. Directed energy deposition (DED) additive manufacturing: Physical characteristics, defects, challenges and applications. *Mater. Today* **2021**, *49*, 271–295. [CrossRef]
12. Babuska, T.F.; Krick, B.A.; Susan, D.F.; Kustas, A.B. Comparison of powder bed fusion and directed energy deposition for tailoring mechanical properties of traditionally brittle alloys. *Manuf. Lett.* **2021**, *28*, 30–34. [CrossRef]
13. Pesode, P.; Barve, S. Additive manufacturing of metallic biomaterials and its biocompatibility. *Mater. Today Proc.* 2022; *in press*. [CrossRef]
14. Wu, M.; Dong, X.; Qu, Y.; Yan, J.; Li, N. Analysis of microstructure and fatigue of cast versus selective laser-melted dental Co-Cr alloy. *J. Prosthet. Dent.* **2022**, *128*, 218.e1–218.e7.
15. Yun, C.S.; Hanawa, T.; Hong, M.H.; Min, B.K.; Kwon, T.Y. Biocompatibility of Ni–Cr alloys, with the same composition, prepared by two new digital manufacturing techniques. *Mater. Lett.* **2021**, *305*, 130761. [CrossRef]
16. Gajera, H.; Shah, D.; Pancholi, N. Effect of SLM process parameters on hardness and microstructure of stainless steel 316 material. *Mater. Today Proc.* **2022**, *50*, 1653–1659.
17. Özsoy, K. Examining mechanical properties of profiles manufactured aluminium extrusion dies using powder bed fusion. *Measurement* **2021**, *177*, 109266. [CrossRef]
18. Korium, M.S.; Roozbahani, H.; Alizadeh, M.; Perepelkina, S.; Handroos, H. Direct metal laser sintering of precious metals for jewelry applications: Process parameter selection and microstructure analysis. *IEEE Access* **2021**, *9*, 126530–126540. [CrossRef]
19. Enneti, R.K.; Morgan, R.; Atre, S.V. Effect of process parameters on the Selective Laser Melting (SLM) of tungsten. *Int. J. Refract. Met. Hard Mater.* **2018**, *71*, 315–319. [CrossRef]
20. Al-Rubaie, K.S.; Melotti, S.; Rabelo, A.; Paiva, J.M.; Elbestawi, M.A.; Veldhuis, S.C. Machinability of SLM-produced Ti6Al4V titanium alloy parts. *J. Manuf. Process.* **2020**, *57*, 768–786. [CrossRef]
21. Pilipović, A.; Brajlih, T.; Drstvenšek, I. Influence of processing parameters on tensile properties of SLS polymer product. *Polymers* **2018**, *10*, 1208. [CrossRef]
22. Idriss, A.I.; Li, J.; Wang, Y.; Guo, Y.; Elfaki, E.A.; Adam, S.A. Selective laser sintering (SLS) and post-processing of prosopis chilensis/polyethersulfone composite (PCPC). *Materials* **2020**, *13*, 3034. [CrossRef]
23. Wu, S.; Yang, L.; Wang, C.; Yan, C.; Shi, Y. Si/SiC ceramic lattices with a triply periodic minimal surface structure prepared by laser powder bed fusion. *Addit. Manuf.* **2022**, *56*, 102910. [CrossRef]

24. Yap, C.Y.; Chua, C.K.; Dong, Z.L.; Liu, Z.H.; Zhang, D.Q.; Loh, L.E.; Sing, S.L. Review of selective laser melting: Materials and applications. *Appl. Phys. Rev.* **2015**, *2*, 041101. [CrossRef]
25. Javaid, M.; Haleem, A. Current status and applications of additive manufacturing in dentistry: A literature-based review. *J. Oral Biol. Craniofac. Res.* **2019**, *9*, 179–185. [CrossRef] [PubMed]
26. Aufa, A.N.; Hassan, M.Z.; Ismail, Z. Recent advances in Ti-6Al-4V additively manufactured by selective laser melting for biomedical implants: Prospect development. *J. Alloys Compd.* **2022**, *896*, 163072. [CrossRef]
27. Gao, B.; Zhao, H.; Peng, L.; Sun, Z. A review of research progress in selective laser melting (SLM). *Micromachines* **2022**, *14*, 57. [CrossRef]
28. Blakey-Milner, B.; Gradl, P.; Snedden, G.; Brooks, M.; Pitot, J.; Lopez, E.; Leary, M.; Berto, F.; Du Plessis, A. Metal additive manufacturing in aerospace: A review. *Mater. Des.* **2021**, *209*, 110008. [CrossRef]
29. Charles, A.; Hofer, A.; Elkaseer, A.; Scholz, S.G. Additive Manufacturing in the Automotive Industry and the Potential for Driving the Green and Electric Transition. In *Sustainable Design and Manufacturing*; Scholz, S.G., Howlett, R.J., Setchi, R., Eds.; Smart Innovation, Systems and Technologies; Springer: Singapore, 2022; Volume 262.
30. Zito, D.; Carlotto, A.; Loggi, A.; Sbornicchia, P.; Maggian, D.; Progold, S.A. Optimization of SLM technology main parameters in the production of gold and platinum jewelry. In Proceedings of the Santa Fe Symposium on Jewelry Manufacturing Technology, Albuquerque, NM, USA, 18–21 May 2014; Met-Chem Research. Volume 47, pp. 439–470.
31. Salmi, M. Additive manufacturing processes in medical applications. *Materials* **2021**, *14*, 191. [CrossRef]
32. Cortis, G.; Mileti, I.; Nalli, F.; Palermo, E.; Cortese, L. Additive manufacturing structural redesign of hip prostheses for stress-shielding reduction and improved functionality and safety. *Mech. Mater.* **2022**, *165*, 104173. [CrossRef]
33. Ni, J.; Liu, F.; Yang, G.; Lee, G.H.; Chung, S.M.; Lee, I.S.; Chen, C. 3D-printed Ti6Al4V femoral component of knee: Improvements in wear and biological properties by AIP TiN and TiCrN coating. *J. Mater. Res. Technol.* **2021**, *14*, 2322–2332. [CrossRef]
34. Sharma, N.; Ostas, D.; Rotar, H.; Brantner, P.; Thieringer, F.M. Design and additive manufacturing of a biomimetic customized cranial implant based on voronoi diagram. *Front. Physiol.* **2021**, *12*, 647923. [CrossRef]
35. Yang, H.J.; Oh, J.H. Reconstruction of Mandibular Contour Defect Using Patient-Specific Titanium Implant Manufactured by Selective Laser Melting Method. *J. Craniofac. Surg.* **2022**, *33*, 2055. [CrossRef]
36. Nahata, S.; Ozdoganlar, O.B. Feasibility of metal additive manufacturing for fabricating custom surgical instrumentation for hip and knee implants. *Procedia Manuf.* **2019**, *34*, 772–779. [CrossRef]
37. Munir, K.; Biesiekierski, A.; Wen, C.; Li, Y. Powder metallurgy in manufacturing of medical devices. In *Metallic Biomaterials Processing and Medical Device Manufacturing*; Woodhead Publishing: Sawston, UK, 2020; pp. 159–190.
38. Oliveira, T.T.; Reis, A.C. Fabrication of dental implants by the additive manufacturing method: A systematic review. *J. Prosthet. Dent.* **2019**, *122*, 270–274. [CrossRef]
39. SLM Solutions, SLM Medical and Dental Applications. Available online: https://www.slm-solutions.com/industries/healthcare/ (accessed on 1 August 2023).
40. Lesyk, D.; Martinez, S.; Mordyuk, B.; Dzhemelinskyi, V.; Lamikiz, A. Surface Finishing of Complexly Shaped Parts Fabricated by Selective Laser Melting. In *Advanced Manufacturing Processes*; Tonkonogyi, V., Ivanov, V., Trojanowska, J., Oborskyi, G., Edl, M., Kuric, I., Pavlenko, I., Dasic, P., Eds.; Lecture Notes in Mechanical Engineering; Springer: Cham, Switzerland, 2019.
41. Boban, J.; Ahmed, A.; Jithinraj, E.K.; Rahman, M.A.; Rahman, M. Polishing of additive manufactured metallic components: Retrospect on existing methods and future prospects. *Int. J. Adv. Manuf. Technol.* **2022**, *121*, 83–125. [CrossRef]
42. Baicheng, Z.; Xiaohua, L.; Jiaming, B.; Junfeng, G.; Pan, W.; Chen-nan, S.; Muiling, N.; Guojun, Q.; Jun, W. Study of selective laser melting (SLM) Inconel 718 part surface improvement by electrochemical polishing. *Mater. Des.* **2017**, *116*, 531–537. [CrossRef]
43. Rashid, H. The effect of surface roughness on ceramics used in dentistry: A review of literature. *Eur. J. Dent.* **2014**, *8*, 571–579. [CrossRef] [PubMed]
44. Velioglu, N.; Akova, T.; Ozkomur, A. Effects of hard thin-film coatings on adhesion of early colonizer bacteria over titanium surfaces. *Implant Dent.* **2016**, *25*, 114–121. [CrossRef]
45. Øilo, M.; Nesse, H.; Lundberg, O.J.; Gjerdet, N.R. Mechanical properties of cobalt-chromium 3-unit fixed dental prostheses fabricated by casting, milling, and additive manufacturing. *J. Prosthet. Dent.* **2018**, *120*, 156-e1. [CrossRef] [PubMed]
46. Zhou, Y.; Li, N.; Yan, J.; Zeng, Q. Comparative analysis of the microstructures and mechanical properties of Co-Cr dental alloys fabricated by different methods. *J. Prosthet. Dent.* **2018**, *120*, 617–623. [CrossRef]
47. Han, X.; Sawada, T.; Schille, C.; Schweizer, E.; Scheideler, L.; Geis-Gerstorfer, J.; Rupp, F.; Spintzyk, S. Comparative analysis of mechanical properties and metal-ceramic bond strength of Co-Cr dental alloy fabricated by different manufacturing processes. *Materials* **2018**, *11*, 1801. [CrossRef]
48. Baciu, E.R.; Cimpoeșu, R.; Vițalariu, A.; Baciu, C.; Cimpoeșu, N.; Sodor, A.; Zegan, G.; Murariu, A. Surface analysis of 3D (SLM) Co–Cr–W dental metallic materials. *Appl. Sci.* **2020**, *11*, 255. [CrossRef]
49. Shu, T.; Zhang, Y.; Sun, G.; Pan, Y.; He, G.; Cheng, Y.; Li, A.; Pei, D. Enhanced osseointegration by the hierarchical micro-nano topography on selective laser melting Ti-6Al-4V dental implants. *Front. Bioeng. Biotechnol.* **2021**, *8*, 621601. [CrossRef]
50. Moraru, E.; Dontu, G.O.; Cananau, S.; Stanescu, V.A. Approaches and Processing Technologies for Medical Devices: Considerations from Micro- and Macroscale Perspectives. In *International Conference on Reliable Systems Engineering (ICoRSE)—2023*; Ciobotă, D.D., Ed.; Lecture Notes in Networks and Systems; Springer: Cham, Switzerland, 2023; Volume 762.

51. Tarcolea, M.; Hancu, V.; Miculescu, F.; Smatrea, O.; Coman, C.; Comaneanu, R.M.; Ormenisan, A. Research on microstructural and chemical inhomogeneity in cast metal crowns made of CoCrMoW alloy. *Rev. Chim.* **2015**, *66*, 1143.
52. Branzei, M.; Pencea, I.; Matei, A.A.; Sfat, C.E.; Antoniac, I.V.; Turcu, R.N.; Manoliu, V. Influence of high temperature exposure on the adhesion of a micro-composite refractory enamel to a Ni–18Cr–12W superalloy. *J. Adhes. Sci. Technol.* **2017**, *31*, 2555–2570. [CrossRef]
53. *ISO 14577-1:2015*; Metallic Materials—Instrumented Indentation Test for Hardness and Materials Parameters—Part 1: Test Method. ISO: Geneve, Switzerland, 2015.
54. Xia, F.; Youcef-Toumi, K. Advanced Atomic Force Microscopy Modes for Biomedical Research. *Biosensors* **2022**, *12*, 1116. [CrossRef] [PubMed]
55. Badita, L.L.; Gheorghe, G.I. Orthopaedic prosthetic systems topographically characterized at nanometric scale. *Adv. Mater. Res.* **2014**, *837*, 675–681. [CrossRef]
56. Evident Olympus, Surface Roughness Measurement. Areal Method (Areal Roughness) Parameters. Available online: https://www.olympus-ims.com/en/metrology/surface-roughness-measurement-portal/ (accessed on 31 August 2023).
57. Keyence, Area Roughness Parameter. Available online: https://www.keyence.com/ss/products/microscope/roughness/surface/parameters.jsp (accessed on 31 August 2023).

materials

MDPI

Article

Microstructural Evolution, Mechanical Properties and Tribological Behavior of B_4C-Reinforced Ti In Situ Composites Produced by Laser Powder Bed Fusion

Jingguang Du [1], Yaojia Ren [1], Xinyan Liu [1,2], Feng Xu [1,2], Xiaoteng Wang [3], Runhua Zhou [4], Ian Baker [5] and Hong Wu [1,*]

1 State Key Laboratory of Powder Metallurgy, Central South University, Changsha 410083, China
2 Farsoon Technologies, Changsha 410205, China
3 Research Institute of Smart Manufacturing, China Railway Construction Heavy Industry Co., Ltd., Changsha 410100, China
4 School of Mechanical and Aerospace Engineering, Nanyang Technological University, Singapore 639798, Singapore
5 Thayer School of Engineering, Dartmouth College, Hanover, NH 03755, USA
* Correspondence: hwucsu@csu.edu.cn

Abstract: Based on the advantage of rapid net-shape fabrication, laser powder bed fusion (LPBF) is utilized to process B_4C-reinforced Ti composites. The effect of volumetric energy density (*VED*) on the relative density, microstructural evolution, tensile properties and wear behaviors of B_4C-reinforced Ti composites were systematically investigated. The LPBF-ed samples with high relative density (>99%) can be achieved, while the pores and un-melted powders can be observed in the sample owing to the low energy input (33 J/mm^3). The additive particulates B_4C were transformed into needle-like TiB whiskers with nano-scale while C dissolved in the Ti matrix. Fine-scale grains (<10 μm) with random crystallographic orientation can be achieved and the residual stress shows a downtrend as the *VED* increases. Through the analysis of the tensile and wear tests, the sample at 61 J/mm^3 *VED* showed a good combination of strength and wear performance, with an ultimate tensile strength of 951 MPa and a wear rate of 3.91×10^{-4} mm^3·N^{-1}m^{-1}. The microstructural evolution in *VED* changes and the corresponding underlying strengthening mechanisms of LPBF-ed Ti + B_4C composites are conducted in detail.

Keywords: laser powder bed fusion; titanium composite; microstructural evolution; mechanical property; tribological behavior

1. Introduction

Titanium and its alloys have the advantages of high specific strength and stiffness combined with superior corrosion resistance, suitable for many industries including aerospace, marine, and automotive. However, the wear resistance of Ti limits the application of the areas mentioned above. To date, much attention has been devoted to titanium matrix composites (TMCs), which utilizes particulate reinforcement technology for strengthening the tribological properties, such as TiB, TiC, and TiN [1–3]. Nevertheless, the limited bond strength between ex situ category reinforcements and the Ti matrix may cause premature failure. As one of the ceramic particulates used, B_4C shows superior stability combined with low density (2.52 g/cm^3) and high hardness (~3000 HV) [4,5]. The introduction of B_4C can efficiently reduce the weight of TMCs and the in situ formed TiB and TiC ceramic-reinforcement phase in TMCs leads to good interfacial bonding, outstanding thermodynamic stability, and fine-scale distribution [6–9]. In addition, the in situ reinforcement phase can provide an improvement in the mechanical properties of TMCs. For instance, Zhang et al. [10] demonstrated that the strength and high-temperature creep behavior of the Ti matrix increased with in situ TiB formation. Yu et al. found that fine in situ TiC particles can be obtained in TiC/Ti coatings processed by induction cladding, which is beneficial to harden the Ti matrix [11].

Many traditional manufacturing processes, such as vacuum casting or powder metallurgy, have been utilized to prepare Ti + B_4C composites [12–14]. However, several issues, e.g., poor wettability, inferior densification response, and cracks, restrict the application of TMCs traditional processes. Besides, coarse carbide and boride particles are formed in TMCs through conventional processing methods owing to the low cooling rate after processing. These brittle phases are the origin of crack initiation and propagation [12,15].

As a promising 3D printing method, laser powder bed fusion (LPBF) can produce complex-shaped parts with almost fully dense structure, which enables to reduce the cost and improves the efficiency of the TMCs process [16–19]. During the LPBF process, a high laser power melts individual layers of material. Transient high temperature (~10^5 K) and ultra-high cooling rate (10^4~10^6 K/s) can be obtained from a high beam scanning speed (500–1500 mm/s), leading to fine microstructures and superior mechanical properties [20–23]. These indicate that LPBF is likely to be an optimal process to produce TMCs with B_4C addition.

Despite the benefits, LPBF fabrication of TMCs is limited by quality defects including pores and cracks due to its complex thermal history [24–26]. To obtain an in-depth understanding of the LPBF processing parameters–microstructure–properties relationship, manipulating the LPBF processing parameters, e.g., laser power and scanning speed, can be used to control the quality of LPBF processed components [27,28]. However, there are numerous LPBF processing parameters, which are interactive. To summarize, volumetric energy density (*VED*) has been used as an engineering parameter to predict the microstructure and properties of LPBF-ed alloys [29]. In the study of Wu et al. [30], the influence of *VED* on microstructural evolution and mechanical behaviors of LPBF-ed AlSi10Mg was investigated. They found that the relative density and yield strength of LPBF-ed AlSi10Mg decreased as *VED* increased from 40 to 90 J/mm^3. Liu et al. systematically studied the effects of the *VED* on LPBF-ed Ti6Al4V and showed that the tensile strength increased and then dropped with increasing *VED* (32.7 to 132.4 J/mm^3) [31]. This indicates that the effect of *VED* on different alloy systems is different. Currently, optimization of the LPBF processing parameters for Ti + B_4C is rarely reported [32]. The influence of *VED* on the microstructure and properties of Ti + B_4C composites needs further exploration. Thus, this paper aims to obtain an understanding of solidification microstructural evolution and the mechanical properties and tribological behavior of these composites.

In the present study, B_4C-reinforced Ti in situ composites produced by LPBF were used as a model to investigate the effect of *VED* on the microstructural evolution, mechanical properties, and tribological behavior. As a parametric study, an optimal LPBF processing parameter was determined to produce Ti composites with high density and enhanced properties. In addition, the corresponding strengthening mechanisms of LPBF-ed Ti + B_4C composites are studied.

2. Experiments

2.1. Powder Preparation

Spherical unalloyed titanium powders (TA1, D10 = 18.4 μm, D50 = 30.6 μm, D90 = 47.1 μm, purchased from TIJO Inc., Changsha, China) and nano-sized B_4C powders (TIJO Inc., Changsha, China) with an average diameter~500 nm was used in this study. The chemical composition of unalloyed Ti powder was detected using an inductively coupled plasma atomic emission spectrometer (ICP-AES), seen in Table 1. The powders were blended under argon in a mixer at room temperature for 12 h with a rotation rate of 20 rpm, and its distribution of particle size was seen in Figure 1a. The B_4C content was 0.5 wt.%. A secondary electron (SE) image of the as-mixed powders is shown in Figure 1b.

Table 1. The chemical composition of unalloyed Ti powder.

Unalloyed Ti Powder	H	C	N	O	Ti
Chemical composition (wt.%)	0.006	0.007	0.023	0.098	Bal.

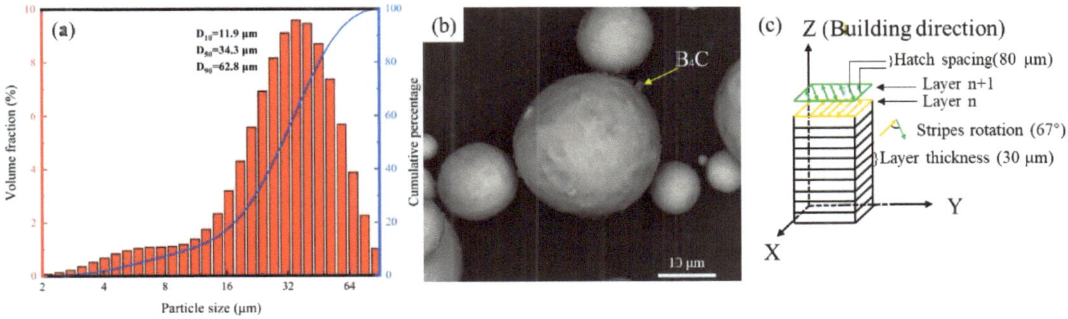

Figure 1. (**a**) Distribution of particle size for the Ti + B_4C mixture, (**b**) SE image of the Ti + B_4C mixture, and (**c**) schematic illustration of the laser scanning strategy.

2.2. LPBF Process

Both cubic (10 × 10 × 10 mm^3) and dogbone-shaped tensile samples (gauge length 9.525 mm, gauge width 2 mm and thickness 2 mm) of the titanium-based composite were produced using an FS121 M LPBF machine equipped with a 500 W fiber laser (Farsoon Inc., Changsha, China). Several samples were built on Ti base plates with area dimensions of 110 mm × 110 mm. Stripes in subsequent layers were rotated by 67° with respect to the previous layer, and sample borders were outlined with a separate perimeter scan. A schematic of the laser scanning strategy is shown in Figure 1c. During the LPBF processing, high-purity argon was used to fill the processing chamber, resulting in less than 100 ppm oxygen present. The *VED* can be calculated using equation (1) from the LPBF parameters used, i.e., the laser power (P, 160–280 W), the scanning speed (v, 1000–2000 mm/s), the hatch distance (h, 80 μm), and the layer thickness (t, 30 μm).

$$VED = \frac{P}{vht} \tag{1}$$

2.3. Microstructural Characterization

For microstructural characterization, specimens were mirror polished using SiC paper and 0.25 μm SiO_2 suspension successively, which are etched using Kroll's reagent (2 mL HF, 6 mL HNO_3, and 92 mL H_2O) for ~10 s. Optical microscopy (OM, LEICA DM4500P, LEICA, Deerfield, IL, USA) was used to characterize the microstructure at low magnification. Phase identification was performed using an X-ray diffractometer (XRD, Advance D8, Bruker, Billerica, MA, USA) with Cu-Kα radiation and a step size of 0.02°. Further microstructural characterization was performed using a scanning electron microscope (SEM, MAIA3 TESCAN, Brno, Czechia) at an accelerating voltage of 20 kV. Texture analysis of the samples was performed using electron backscattered diffraction (EBSD) at a step size of 0.7 μm. Transmission electron microscope (TEM, FEI Titan G2, FEI, Hillsboro, OR, USA) analysis was performed at an accelerating voltage of 200 kV, and the compositions were determined using energy dispersive spectrometer (EDS).

2.4. Mechanical and Tribological Tests

The physical and mechanical properties of the LPBF-ed specimens were investigated via density measurements, nanoindentation measurements, tensile tests, and tribological studies. The density (ρ) was determined using the Archimedes method from:

$$\rho = \frac{W_a \times \rho_w}{W_a - W} \quad (2)$$

where W_a and W are the weights of the sample in air and water, respectively, and ρ_w is the density of the water. The density machine is Sartorius MSA324S-000-DU and each reported value is the average of three measurements. The theoretical density of Ti composite is 4.486 g/cm^3. After analyzing the relative density results (Figure 2), the three specimens (named S1, S2, and S3) with *VED* values ranging from 33 to 117 J/mm^3 are selected and discussed in this paper: the processing parameters associated with these specimens are listed in Table 2.

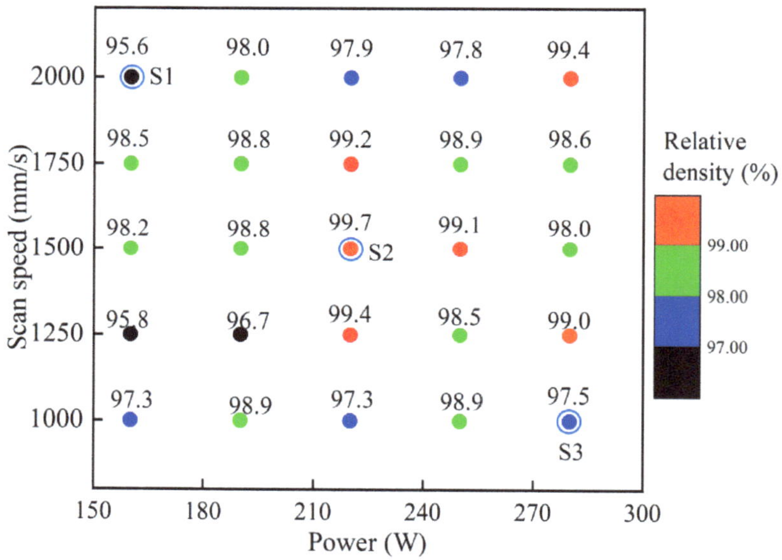

Figure 2. Relative density of LPBF-ed Ti + B_4C composite.

Table 2. LPBF processing parameters for Ti + B_4C composite specimens.

Specimen	P (W)	v (mm/s)	h (μm)	t (μm)	*VED* (J/mm^3)
S1	160	2000	30	80	33
S2	220	1500	30	80	61
S3	280	1000	30	80	117

The nanoindentation measurements were undertaken using a MCT + UNHT machine (CSM Company, Kaisten, Switzerland) at room temperature. The load was set at 30 mN load with a loading time of 15 s and a Berkvoich indenter was used. Each specimen was measured at least 3 times. The reduced Young's modulus is defined as:

$$\frac{1}{E_r} = \frac{1 - v^2}{E} + \frac{1 - v_i^2}{E_i} \quad (3)$$

where E and v are the elastic modulus and Poisson's ratio for the specimen, while E_i and v_i are the corresponding values for the indenter.

Tensile tests were performed at room temperature at an initial strain rate of $1 \times 10^{-3}s^{-1}$ on specimens cut using a wire electrical discharge machine perpendicular to the building direction and ground to a surface finish using 3000 grid SiC paper. The strain was measured with a video probe. Dry sliding wear tests were performed on specimens' flat ground. A 4 mm diameter Si_3Ni_4 ball was used as the counter-material. The load, rotation speed, rotation radius, and time were at 10 N, 10 Hz, 1 mm, and 30 min, respectively. The tensile fracture surfaces and worn surfaces were examined using the secondary electron mode in the SEM.

3. Results

3.1. Microstructure

Figure 3 shows typical OM images of S1, S2 and S3 viewed in the x–y plane. The melt track, also known as the melt pool, can be clearly observed. It is evident that the rotation angle between melt tracks is about 67°, as expected. Micropores were presented in S1 resulting in lower relative density (~96%), which may be attributed to insufficient energy penetration at the lowest *VED*. The morphology of S2 was similar to that of S3. The excellent track-to-track bonding suggested that the alloys with a near-dense structure show good mechanical properties.

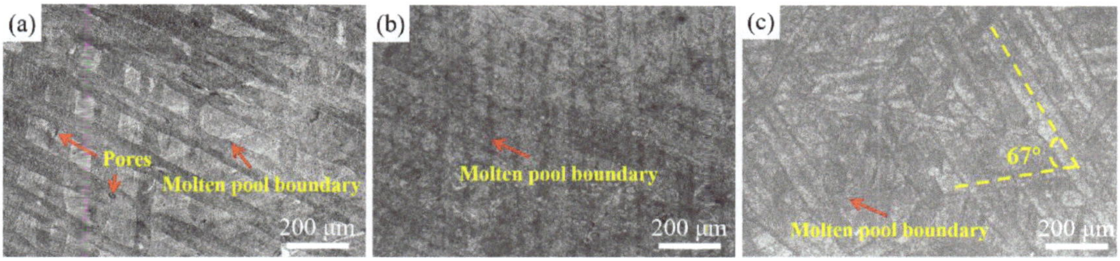

Figure 3. OM images of (**a**) S1, (**b**) S2, and (**c**) S3.

Figure 4 shows the surface morphologies of S1, S2 and S3. There were both un-melted powders and island-shape phenomenon on the surface of S1 (Figure 4a), arising from the limited energy penetration at a *VED* of 33 J/mm^3. As the *VED* increased to 61 J/mm^3 (specimen S2), both the number of un-melted powders and the surface island-shape phenomenon decreased substantially. When the *VED* further increased to 117 J/mm^3 (specimen S3), the surface became smoother and free of any evident defects, a result that can be attributed to the greater laser power penetrating the powder-bed [33].

Figure 4d–f show typical microstructures of LPBF-ed samples in the x–y plane. The *VED* exerts a significant influence on the microstructure of the Ti + B_4C composites. Unlike the other specimens, S1 exhibited cellular microstructure As the *VED* increased from 61 J/mm^3 to 117 J/mm^3, a change to a columnar structure can be observed. The difference in microstructure is dependent on the complex solidification behaviors in the molten pool caused by various *VED*, which is discussed in Section 4.1. In addition, needle-like TiB whiskers of different sizes can be seen in all three samples. Clusters composed of TiB whiskers were present in the LPBF-ed microstructure, where parallel whiskers were stuck to each other. The TiB whiskers grew into this morphology due to the faster growth rate in the [010] axis than that in other directions [34].

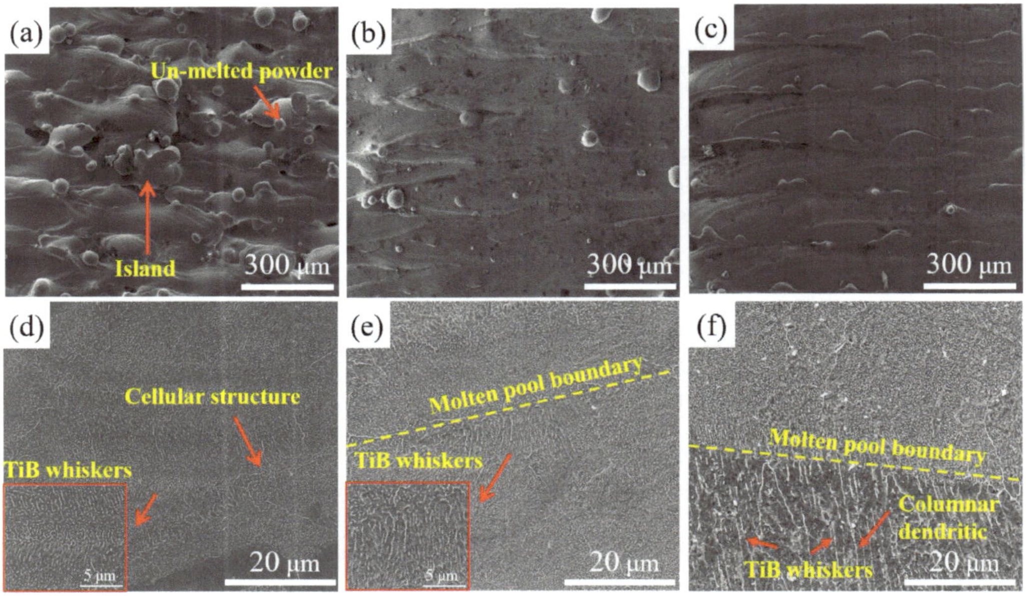

Figure 4. SE images of surface and microstructure of LPBF-ed Ti + B_4C in situ composites: (**a**,**d**) S1; (**b**,**e**) S2; (**c**,**f**) S3.

3.2. EBSD Characterization

The effect of *VED* on the grain size and crystallographic orientations of the LPBF-ed B_4C-reinforced Ti in situ composites was studied using EBSD. Inverse pole figures (IPF) viewed perpendicular to the building direction (BD) of S1, S2, and S3 are shown in Figure 5. The grains in the three samples were mostly smaller than 10 μm, while a few more coarse grains were observed in S3. The average α lath width of three samples was determined to be 0.67 μm, 0.75 μm, and 0.89 μm, respectively. The small increase in lath width can be attributed to the greater heat input from the increased *VED*. Besides, the top-viewed crystallographic orientations of three specimens in Figure 5 show crystallographic directions were random.

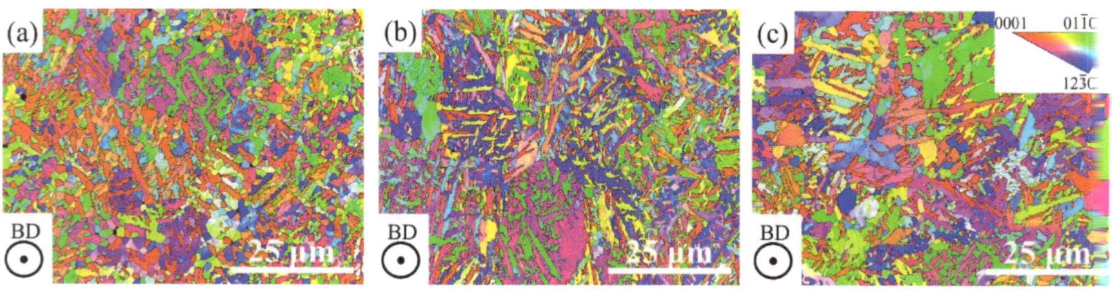

Figure 5. IPFs perpendicular to BD of samples with different *VED*: (**a**) S1; (**b**) S2; (**c**) S3.

3.3. TEM Analysis

To further clarify the microstructure of the LPBF-ed Ti + B_4C composites, specimen S2 was examined in the TEM. Figure 6a shows that the needle-like whiskers with sizes ranging from 50 to 100 nm were randomly distributed in the matrix. However, TiC and TiB are hardly observed in TEM-EDS maps. In α-Ti alloys, the solid solubility of C is 0.458 wt.% at 1173K and drops to 0.126 wt.% at 873K [35]. During the LPBF process, 0.104 wt.% of C dissolved into S2, where the carbon content is below the maximum carbon solubility. This is the reason that TiC particles are not observed at room temperature.

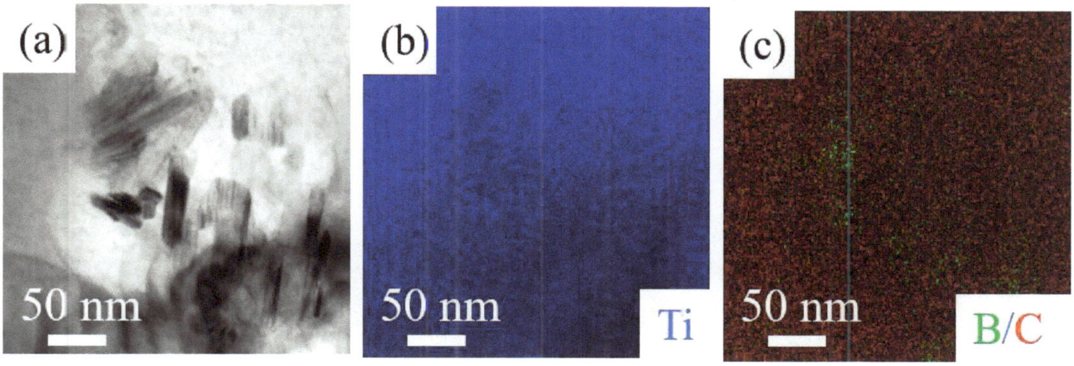

Figure 6. (**a**) BFTEM images of LPBF-ed S2; (**b**,**c**) corresponding EDS map showing the Ti and B/C distribution.

High-resolution TEM (HRTEM) was performed to examine the interface between the TiB whisker and the α-Ti matrix. Figure 7a shows a clean interface between the matrix and the reinforcing phase, illustrating good bonding between the TiB whisker and the α-Ti. The α-Ti matrix and TiB whiskers were identified via determining the lattice parameters to be 0.159 nm and 0.250 nm, respectively. However, the theoretical lattice spacing of $\alpha\text{-Ti}_{d\{110\}}$ is known to be 0.147 nm, which is smaller than the experimental results. Kværndrup et al. reported that the interstitial atoms (O, N, or H) dissolution into α-Ti leads to the increase of lattice parameter [36]. The results indicated that C is dissolved into Ti, leading to lattice spacing expansion. The corresponding fast Fourier transform (FFT) patterns of the LPBF-ed Ti + B_4C composite are shown in Figure 7b,c, which further confirms the existence of the TiB whisker and the α-Ti phase. In addition, intense streaking in the direction of TiB (110) was observed, demonstrating the existence of stacking faults (SF) in the TiB (110) plane. Kooi et al. reported a similar structure in laser-clad Ti-TiB [34].

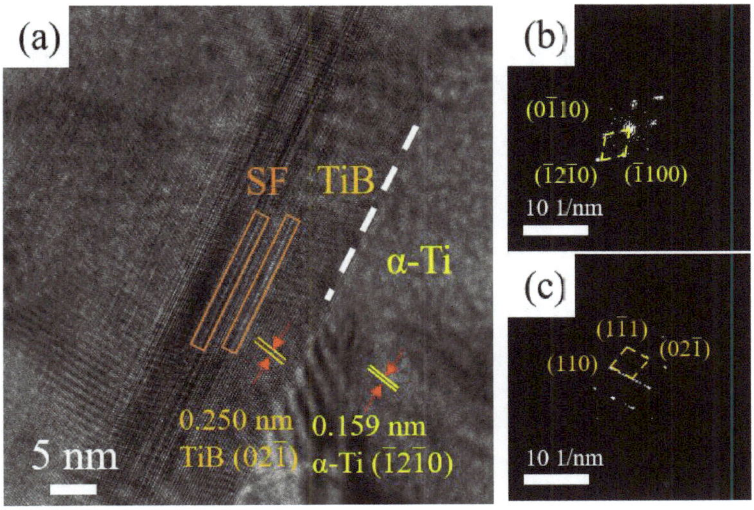

Figure 7. (**a**) HRTEM images of TiB whisker and the Ti matrix; (**b**,**c**) FFT patterns of α-Ti and TiB whisker, respectively.

3.4. Phase Identification

XRD patterns from S1, S2 and S3 are shown in Figure 8. Strong diffraction peaks from hexagonal Ti (HCP) and weak diffraction peaks from the small volume fraction of TiB were found. According to the Ti-B-C ternary phase diagram [37], an in situ reaction occurred during the solidification processing of LPBF-ed samples:

$$Ti + B_4C \rightarrow \beta\text{-}Ti + TiB \rightarrow \alpha\text{-}Ti + TiB \quad (4)$$

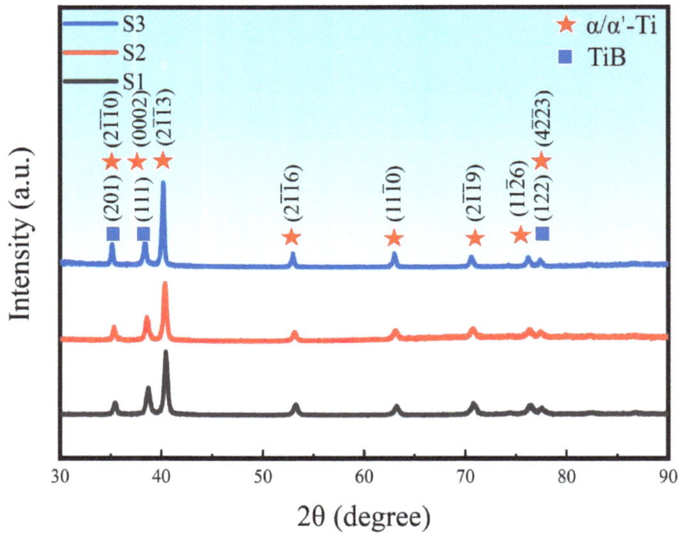

Figure 8. XRD results of three LPBF-ed samples with various *VED*s.

3.5. Mechanical Properties

Tensile tests were conducted to evaluate the LPBF-ed B_4C-reinforced Ti in situ composites and their representative engineering stress-engineering strain responses are displayed in Figure 9: values for the yield strength (YS), ultimate tensile strength (UTS), elongation (ε) is summarized in Table 3. S1 had the lowest relative density (95.6%) and exhibited the lowest ε, but had the highest YS at 768 $\pm$ 10 MPa. Compared with S1, the YS value of S2 was ~30 MPa less at 738 $\pm$ 6 MPa, but its ε increased from 1.7 $\pm$ 0.4% to 6.3 $\pm$ 1.1%, presumably due to a lower number of defects. Compared with S2, S3 had the similar ε at 7.4 $\pm$ 1.9%. However, S3 had a lower YS value (664 $\pm$ 9 MPa) than those of the other two specimens. As increasing *VED*, the pore morphology in Ti alloys changed from irregular shape to near-spherical shapes [31]. During tensile test, samples with spherical pores have a more uniform distribution of strain, which may lead to better ductility performance than that of samples with irregular pores. The decrease of YS can be attributed to the coarse grain in S3 caused by large energy input. As shown in Table 3, the introduction of B_4C to pure Ti can achieve high tensile strength and reasonable elongation with comparison of pure Ti produced by LPBF, wrought or cast [38–42].

Nanoindentation tests were conducted on the three specimens and the smooth load-displacement curves obtained are shown in Figure 10a. It can be observed that the indentation depth of the LPBF-ed Ti + B_4C composites becomes slightly deeper with increasing *VED*. Figure 10b is a histogram of the hardness (*H*) and reduced Young's modulus, (*Er*) for the three samples. The increasing *VED* led to a slight decrease in *H* from 4.6 GPa to 3.8 GPa. The reduced Young's moduli of S1 and S2 were similar (~128 GPa) but were ~8% greater than that of S3.

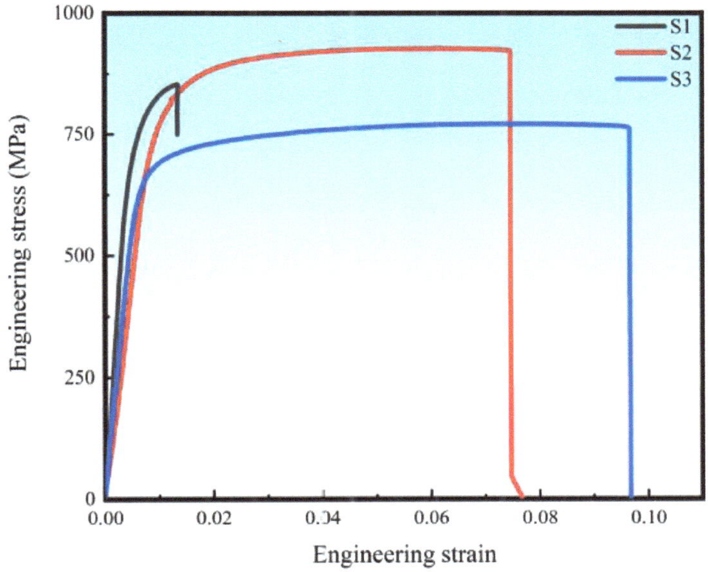

Figure 9. Engineering strain-stress curves of LPBF-ed B_4C-reinforced Ti in situ composites.

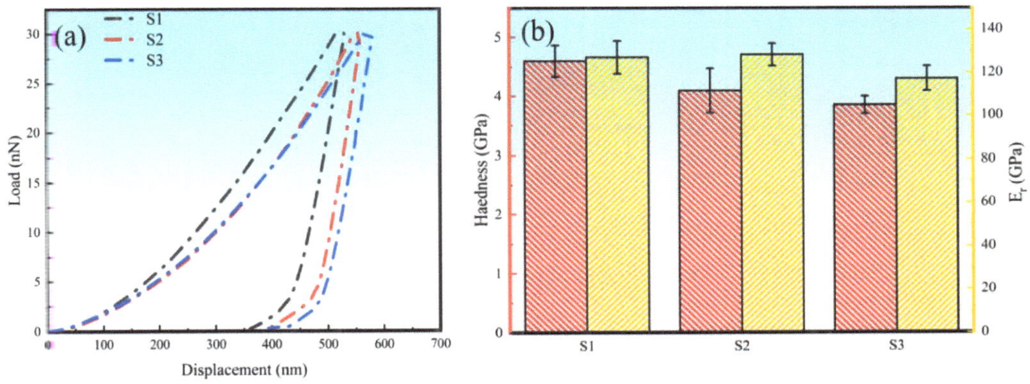

Figure 10. (**a**) Displacement-load response and (**b**) hardness and reduced Young's modulus results of three samples.

Table 3. Summary of the relative density and average mechanical properties.

Sample	Condition	YS (MPa)	UTS (MPa)	ε (%)	Reference
S1	LPBF	768	821	1.7	This work
S2	LPBF	738	951	6.3	This work
S3	LPBF	664	771	7.4	This work
Ti	LPBF	521	607	10.4	[38]
Ti	LPBF	590	665	19	[39]
Ti	LPBF	407	469	14.7	[40]
Ti	LPBF	420	510	18	[41]
Ti	Wrought	317	481	28.9	[38]
Ti	Cast	351	466	30	[42]

3.6. Tribological Behavior

Figure 11 presents the friction coefficient and wear rate from wear tests of the LPBF-ed Ti + B_4C composites. The wear rate coefficient (K_c) was calculated from (5):

$$K_c = \frac{2\pi rA}{FS} \quad (5)$$

where r and A are the radius and area of wear tracks, respectively, and F and S are the load and sliding distance, respectively. The friction coefficient curves of the three samples experienced a short fluctuation within an unstable period (about 3 min) and then became more stable. The average friction coefficients of S1, S2, and S3 were calculated during the steady period (3-30 min). S1 exhibited both a high friction coefficient (0.25 $\pm$ 0.01) and a high K_c (6.2 $\pm$ 2.1 $\times$ 10^{-4} $mm^3 \cdot N^{-1}m^{-1}$). A similar friction coefficient (0.25 $\pm$ 0.01), but the lowest K_c (3.9 $\pm$ 0.6 $\times$ 10^{-4} $mm^3 \cdot N^{-1}m^{-1}$) was obtained for S2. Interestingly, S3 exhibited the lowest friction coefficient (0.22 $\pm$ 0.01) but the highest K_c (12.4 $\pm$ 1.5 $mm^3 \cdot N^{-1}m^{-1}$).

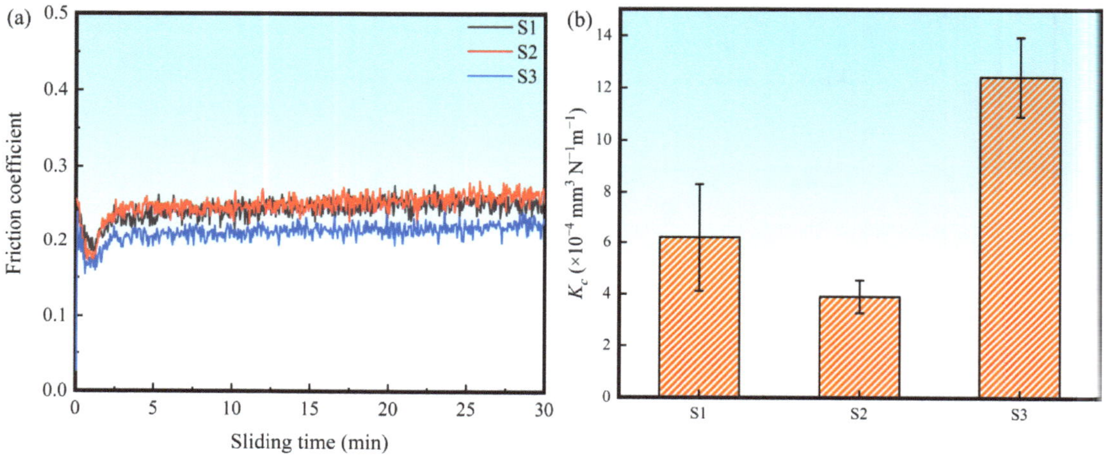

Figure 11. (**a**) Friction coefficient versus time, and (**b**) K_c for the three samples.

The worn surfaces (typical morphologies and 3D surface profiles) and corresponding EDS mapping of S1, S2, and S3 are presented in Figure 12. The shallower grooves along the sliding distance on S1 and S2 surfaces were clear while the severer grooved scratches and more debris can be observed on the S3 surface. Based on the EDS mapping result, it was demonstrated that wear debris were tribological oxides and the tribological layers were composed of oxides. In addition, white debris detected in Figure 12 were considered an oxidative wear character because of the exposure to the air during the dry sliding wear test [43,44].

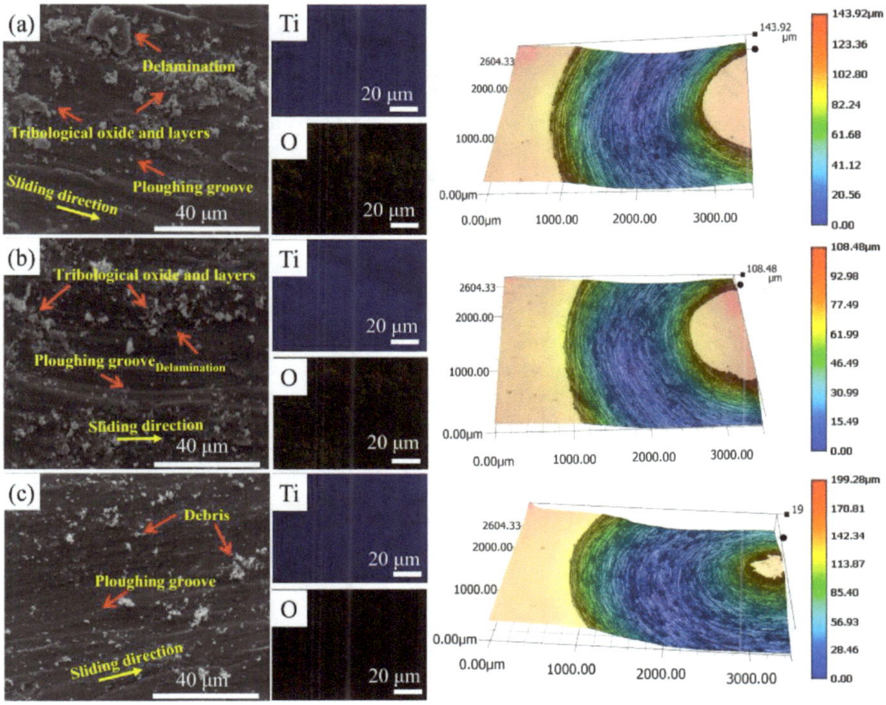

Figure 12. SE images of worn surfaces, corresponding EDS maps and 3D surface profiles: (**a**) S1, (**b**) S2, and (**c**) S3.

4. Discussion

4.1. Microstructural Evolution

When the laser penetrates the powder-bed, the Ti powders first melted and then reacted with the B_4C particles in the molten pool. However, inadequate penetration led to un-melted powder and island-shape phenomenon, which produced a rough surface, as observed in specimen S1. With increasing *VED*, the composite powders could be fully melted and the surface became smooth. In general, the viscosity of the fluid is strongly determined by the temperature in the molten pool, which has a significant influence on surface quality. According to Takamichi [45], the viscosity of fluid (μ) in the molten pool can be calculated from (6):

$$\mu = \alpha\sqrt{\frac{m}{kT}}\gamma \tag{6}$$

where α is a constant, m is the atomic mass, k is the Boltzmann constant, γ is the surface tension of fluid and T is fluid temperature. The surface tension of fluid dropped with increasing fluid temperature [46]. Therefore, the viscosity of fluid decreased caused by a higher *VED*. The fluid in the molten pool spreads easily and a high-quality surface without defects was obtained for higher *VEDs*.

Characteristic solidification microstructures in Figure 4 were formed in the LPBF process. When the laser interacted with the powder bed, B_4C is prone to react with Ti and transforms into in situ formed TiB whiskers. TiB whiskers are melted into the liquid over its melting point (2473 K) or partially dissolved in the region below 2473 K. During the solidification process, the solute atoms were ejected at the solid/liquid interface front, which led to a higher concentration of solute. The retained TiB whisker was captured to act as grain boundaries and eutectic TiB whiskers solidified between the Ti grains. Besides, the length of TiB whiskers became larger, which may be caused by longer solidification time as *VED* increased. The solidification morphology is determined by the ratio of the temperature gradient (G) to solidification rate (R) while G×R governs the microstructural scale. G is given by the temperature field caused by the laser and R is related to beam velocity and the melt pool shape. A change from cellular to columnar structures occurred due to differences in G and R in the molten pool caused by different *VEDs*. The metastable cell structure of S1 formed in the LPBF process can be attributed to the moderate G/R. At the higher *VED* in S3, a lower G/R was obtained, which led to columnar dendrites [47]. Therefore, it was seen that increasing *VED* may lead to a reduced G/R, which changes the microstructure.

4.2. Mechanical Properties Analysis

The increment in the strength of LPBF-ed Ti + B_4C in situ composites can be attributed to three strengthening mechanisms: (1) grain refinement caused by the heterogeneous nucleation on TiB; (2) TiB whiskers acting as reinforcements; and (3) solid-solution strengthening from dissolved carbon. For a better understanding of the strengthening mechanisms, S2 was chosen to quantitatively estimate these effects.

Upon the addition of the B_4C powders, the grain size decreased compared to LPBF-ed Ti [41]. Based on the Hall–Petch relationship, the increase in yield strength by grain size reduction is expressed as:

$$\Delta\sigma_{\text{H-P}} = K\left(d_1^{-\frac{1}{2}} - d_2^{-\frac{1}{2}}\right) \tag{7}$$

where K is a constant as 328 MPa $\mu m^{1/2}$ [48], d_1 and d_2 are α-Ti grain size of the Ti + B_4C composite and the titanium matrix, respectively.

Because of the limited solid solubility in the matrix (~0.02 wt.%), most of the boron was presented in the TiB whiskers. The fiber strengthening effect from the TiB whiskers can be calculated from:

$$\Delta\sigma_{\text{TiB}} = 0.5\sigma_{\text{YSm}} V_{\text{TiB}} \frac{l}{d} \omega_0 \tag{8}$$

where σ_{YSm} is the yield strength of the Ti matrix, V_{TiB}, l/d, and ω_0 are the volume fraction, aspect ratio, and whisker orientation factor for the TiB whiskers, respectively. The orientation of TiB whiskers is random and hence ω_0 = 0.27 [49–51].

The solid solution strengthening from carbon in the Ti matrix can be expressed as [49]:

$$\Delta\sigma_S = \frac{1}{\sqrt{3}} m_T \frac{1}{2(1+v)} E_r \eta^{\frac{3}{2}} c^{\frac{1}{2}} \tag{9}$$

where m_T, v, E_r are the Taylor factor, Poisson's ratio, and elastic modulus of the Ti alloy, respectively. η is related to the change in the lattice constants with carbon concentration in the Ti matrix [52] and c is the atom fraction of carbon (~0.43 wt.%).

Therefore, the theoretical YS of S2 can be calculated from:

$$\sigma_{YS} = \sigma_{\text{YSm}} + \Delta\sigma_{\text{H-P}} + \Delta\sigma_{\text{TiB}} + \Delta\sigma_S \tag{10}$$

Using the Equations (7)–(10), the YS of S2 was calculated to be 698 MPa with parameters listed in Table 4, which agreed relatively well with the experimental value of 712 MPa. Notably, $\Delta\sigma_{\text{H-P}}$, $\Delta\sigma_{\text{TiB}}$, and $\Delta\sigma_S$ are 134 MPa, 22 MPa, and 136 MPa, respectively. The largest increase in strength was from solution strengthening, accounting for 48%. Note that the tensile strength of S3 was significantly lower than that of others, which may be attributed to higher energy penetration resulting in coarser grains.

Table 4. Parameters value for calculation.

Parameter	Value	Reference
K	328 MPa $\mu m^{1/2}$	[48]
d_1	0.75 μm	Measured
d_2	1.8 μm	[41]
σ_{YSm}	420 MPa	[41]
l/d	20	Measured
m_T	3.16	Measured
ω_0	0.27	[53]
E_r	128 GPa	Measured
η	0.08	[49]
v	0.27	[49]

Secondary electron images of the fracture surfaces of the LPBF-ed Ti + B_4C composites are shown in Figure 13. There were several pores and cleavage steps evident in S1 caused by insufficient *VED* (Figure 13a). During tensile testing, crack propagation may begin at such defects, leading to worse ductile performance [54,55]. Such pores were not evident for S2 and S3, which explained the higher ductility for these two specimens. A combination of cleavage steps and dimples can be observed in Figure 13b, which demonstrated that the fracture mechanism was a mixture of brittle fracture and plastic deformation. However, the dimples were small and shallow, compared with that in Figure 13c. This phenomenon explained the plastic deformation in S2 was limited. The main fracture mechanism of S3 was plastic deformation because of the full dimples. In summary, the mechanism of fracture changed from brittleness to ductile failure with the enhanced *VED*.

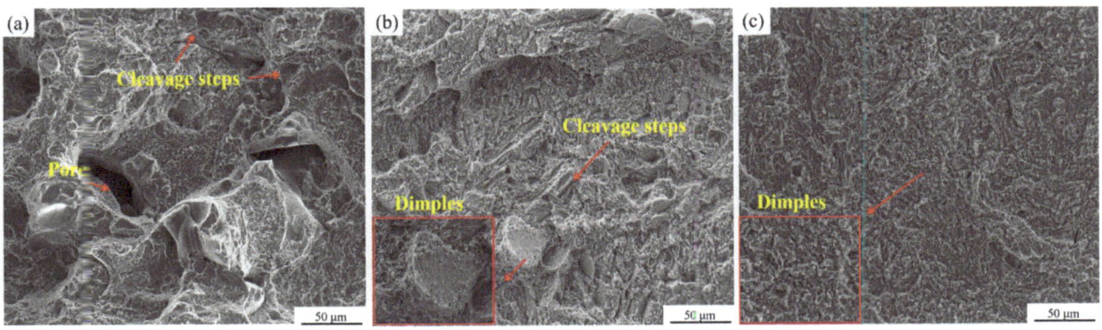

Figure 13. SE images of the fracture surfaces of (**a**) S1, (**b**) S2, and (**c**) S3.

4.3. Tribological Behavior Analysis

Different wear mechanisms of the LPBF-ed samples led to different tribological behavior. The shallow grooves along the sliding distance and tribological layers of S1 and S2 indicated that adhesive and abrasive wear mechanisms predominate during the wear test. The dominant wear mechanism of S3 was abrasive wear, demonstrated by severer grooved scratches and more white debris. The contact and relative sliding between materials and Si_3N_4 generated high tribological heat and local pressure, leading to the formation of tribological oxides and layers on the composite surface. During a long period of high-stress contact, cracks and delamination occurred on the tribological layers. In S1 and S2, tribological layers composed of wear debris were formed due to the applied stress during the wear test. Layers composed of tribological oxide existing on the sample surface can efficiently reduce wear rate and improve wear resistance [44]. In addition, the hardness and strength of materials were considered important indices judging wear resistance [56]. It can be inferred that tribological layers composed of oxide and better strength of S1 and S2 led to better tribological performance than that of S3 (12.4 ± 1.5 $mm^3 \cdot N^{-1} m^{-1}$). While the lower relative density of S1 led to an unstable tribological performance ($6.2 \pm 2.1 \times 10^{-4}$ $mm^3 \cdot N^{-1} m^{-1}$), which may be due to un-melted powders and pores acting as crack initiation [57].

5. Conclusions

In situ Ti composites reinforced by B_4C were successfully processed by LPBF. The effect of *VED* on relative density, microstructure, tensile properties, and wear behaviors have been investigated. The following conclusions are:

(1) A relative density higher than 99% was achieved in S2 and S3. The island-shape phenomenon and un-melted powders in S1 can be attributed to the lower viscosity of fluid caused by inadequate lower energy input.

(2) The microstructural evolution was influenced by *VED*. The cellular microstructure changes gradually to columnar morphology with *VED* increment. A small increase of grain size due to more heat input by *VED* and the crystallographic orientations are random

(3) A superior UTS of 951 ± 21 MPa combined with reasonable ductility (6.3 ± 1.1%) was achieved in the S2. The improved strength was attributed to grain refinement strengthening caused by TiB whiskers, TiB reinforcement mechanism, and solution strengthening of carbon in the Ti matrix. The fracture mechanism changed gradually from brittleness to ductile failure with the *VED* increases.

(4) A low wear rate and friction coefficient of S2 were reached 3.91×10^{-4} $mm^3 \cdot N^{-1} m^{-1}$ and 0.252, respectively. Owing to tribological layers on the sample surface and better strength, S1 and S2 had a better wear behavior than that of S3.

Author Contributions: J.D., conceptualization, data curation, methodology, and writing—original draft. Y.R., conceptualization, formal analysis, investigation, and methodology. X.L., conceptualization and writing—review and editing. F.X., conceptualization. X.W., writing—review and editing. R.Z., conceptualization and methodology. I.B., conceptualization and methodology. H.W., funding acquisition and writing—review and editing. All authors have read and agreed to the published version of the manuscript.

Funding: This research was funded by the National Natural Science Foundation of China (No. 52071346), the Natural Science Foundation of Hunan Province for Distinguished Young Scholars (No. 2023JJ10075), and Central South University Research Programme of Advanced Interdisciplinary Studies (No. 2023QYJC038).

Institutional Review Board Statement: Not applicable.

Informed Consent Statement: Not applicable.

Data Availability Statement: The data presented in this study are available on request from the corresponding author.

Acknowledgments: The authors would thank Sinoma Institute of Materials Research (Guang Zhou) Co., Ltd. for the assistance with the TEM characterization.

Conflicts of Interest: The authors declare no conflict of interest.

References

1. Zhang, J.; Song, B.; Cai, C.; Zhang, L.; Shi, Y. Tailorable microstructure and mechanical properties of selective laser melted TiB/Ti–6Al–4V composite by heat treatment. *Adv. Powder Mater.* **2022**, *1*, 100010. [CrossRef]
2. Zhang, Z.; Chen, J.; He, G.; Yang, G. Fatigue and Mechanical Behavior of Ti-6Al-4V Alloy with CrN and TiN Coating Deposited by Magnetic Filtered Cathodic Vacuum Arc Process. *Coatings* **2019**, *9*, 689. [CrossRef]
3. Ali, T.; Wang, L.; Cheng, X.; Gu, D.; Zhou, Z.; Min, X. The effect of TiC on microstructure and mechanical properties of Ti-5553 beta phase titanium alloy. *Mater. Des.* **2022**, *214*, 110395. [CrossRef]
4. Zhang, W. A review of tribological properties for boron carbide ceramics. *Prog. Mater. Sci.* **2021**, *116*, 100718. [CrossRef]
5. Chand, S.; Chandrasekhar, P.; Sarangi, R.; Nayak, R. Influence of B4C particles on processing and strengthening mechanism in aluminum metal matrix composites-a review. *Mater. Today Proc.* **2019**, *18*, 5356–5363. [CrossRef]
6. Soy, U.; Demir, A.; Findik, F. Friction and wear behaviors of Al-SiC-B4C composites produced by pressure infiltration method. *Ind. Lubr. Tribol.* **2011**, *63*, 387–393. [CrossRef]
7. Sousa, L.; Alves, A.; Guedes, A.; Toptan, F. Corrosion and tribocorrosion behaviour of Ti-B4C composites processed by conventional sintering and hot-pressing technique. *J. Alloys Compd.* **2021**, *885*, 161109. [CrossRef]

8. Han, C.; Babicheva, R.; Chua, J.D.Q.; Ramamurty, U.; Tor, S.B.; Sun, C.-N.; Zhou, K. Microstructure and mechanical properties of (TiB+ TiC)/Ti composites fabricated in situ via selective laser melting of Ti and B4C powders. *Addit. Manuf.* **2020**, *36*, 101466. [CrossRef]
9. Ren, L.; Wu, H.; Agbedor, S.-O.; Lu, Y.; Zhang, Y.; Fang, Q.; Li, J.; Tian, Y.; Baker, I. Microstructure, mechanical and tribological properties of a Ti-5Cu alloy and a B4C/Ti-5Cu in situ composite fabricated by laser powder bed fusion. *Mater. Charact.* **2022**, *192*, 112217. [CrossRef]
10. Zhang, Z.; Qin, J.; Zhang, Z.; Chen, Y.; Lu, W.; Zhang, D. Microstructure effect on mechanical properties of in situ synthesized titanium matrix composites reinforced with TiB and La_2O_3. *Mater. Lett.* **2010**, *64*, 361–363. [CrossRef]
11. Yu, H.L.; Zhang, W.; Wang, H.M.; Ji, X.C.; Song, Z.Y.; Li, X.Y.; Xu, B.S. In-situ synthesis of TiC/Ti composite coating by high frequency induction cladding. *J. Alloys Compd.* **2017**, *701*, 244–255. [CrossRef]
12. Li, S.; Kondoh, K.; Imai, H.; Chen, B.; Jia, L.; Umeda, J.; Fu, Y. Strengthening behavior of in situ-synthesized (TiC–TiB)/Ti composites by powder metallurgy and hot extrusion. *Mater. Des.* **2016**, *95*, 127–132. [CrossRef]
13. Zhang, M.; Yuan, T.; Li, R. Effect of spark plasma sintering on microstructure and friction characteristics of boron carbide. *J. Micromechanics Mol. Phys.* **2018**, *3*, 1750014. [CrossRef]
14. Sabahi Namini, A.; Delbari, S.A.; Nayebi, B.; Shahedi Asl, M.; Parvizi, S. Effect of B4C content on sintering behavior, microstructure and mechanical properties of Ti-based composites fabricated via spark plasma sintering. *Mater. Chem. Phys.* **2020**, *251*, 123087. [CrossRef]
15. Banoth, R.; Sarkar, R.; Bhattacharjee, A.; Nandy, T.K.; Nageswara Rao, G.V.S. Effect of boron and carbon addition on microstructure and mechanical properties of metastable beta titanium alloys. *Mater. Des.* **2015**, *67*, 50–63. [CrossRef]
16. Sun, F.E.; Zhang, B.; Qu, X. High strength Al alloy development for laser powder bed fusion. *J. Micromechanics Mol. Phys.* **2021**, *6*, 214001. [CrossRef]
17. Zhang, T.; Liu, C.-T. Design of titanium alloys by additive manufacturing: A critical review. *Adv. Powder Mater.* **2021**, *1*, 100014. [CrossRef]
18. Chowdhury, S.; Yadaiah, N.; Prakash, C.; Ramakrishna, S.; Dixit, S.; Gupta, L.R.; Buddhi, D. Laser powder bed fusion: A state-of-the-art review of the technology, materials, properties & defects, and numerical modelling. *J. Mater. Res. Technol.* **2022**, *20*, 2109–2172. [CrossRef]
19. Luo, X.; Song, T.; Wang, F.; Lu, H.; Kang, L.; Ma, H.; Li, D.; Gebert, A.; Yang, C. Phase selection-oriented mechanical properties tailoring for β-type TiNbZrTaSi alloy fabricated by laser powder bed fusion. *Adv. Powder Mater.* **2023**, *2*, 100118. [CrossRef]
20. Gu, D.; Shi, X.; Poprawe, R.; Bourell, D.L.; Setchi, R.; Zhu, J. Material-structure-performance integrated laser-metal additive manufacturing. *Science* **2021**, *372*, eabg1487. [CrossRef]
21. Zhang, D.; Qiu, D.; Gibson, M.A.; Zheng, Y.; Fraser, H.L.; StJohn, D.H.; Easton, M.A. Additive manufacturing of ultrafine-grained high-strength titanium alloys. *Nature* **2019**, *576*, 91–95. [CrossRef] [PubMed]
22. Yadroitsev, I.; Krakhmalev, P.; Yadroitsava, I. Titanium alloys manufactured by in situ alloying during laser powder bed fusion. *JOM* **2017**, *69*, 2725–2730. [CrossRef]
23. Tan, C.; Ma, W.; Deng, C.; Zhang, D.; Zhou, K. Additive manufacturing SiC-reinforced maraging steel: Parameter optimisation, microstructure and properties. *Adv. Powder Mater.* **2023**, *2*, 100076. [CrossRef]
24. Yavari, R.; Smoqi, Z.; Riensche, A.; Bevans, B.; Kobir, H.; Mendoza, H.; Song, H.; Cole, K.; Rao, P. Part-scale thermal simulation of laser powder bed fusion using graph theory: Effect of thermal history on porosity, microstructure evolution, and recoater crash. *Mater. Des.* **2021**, *204*, 109685. [CrossRef]
25. Zhang, J.; Yuan, W.; Song, B.; Yin, S.; Wang, X.; Wei, Q.; Shi, Y. Towards understanding metallurgical defect formation of selective laser melted wrought aluminum alloys. *Adv. Powder Mater.* **2022**, *1*, 100035. [CrossRef]
26. Wang, D.; Yang, Y.; Wang, Y.; Yang, L.; Wang, H.; Yang, S. Introduction to the Special Issue on Design and Simulation in AdditiveManufacturing. *Comput. Model. Eng. Sci.* **2021**, *126*, 1–4. [CrossRef]
27. Huang, S.; Narayan, R.L.; Tan, J.H.K.; Sing, S.L.; Yeong, W.Y. Resolving the porosity-unmelted inclusion dilemma during in-situ alloying of Ti34Nb via laser powder bed fusion. *Acta Mater.* **2021**, *204*, 116522. [CrossRef]
28. Wu, H.; Ren, J.; Huang, Q.; Zai, X.; Liu, L.; Chen, C.; Liu, S.; Yang, X.; Li, R. Effect of laser parameters on microstructure, metallurgical defects and property of AlSi10Mg printed by selective laser melting. *J. Micromechanics Mol. Phys.* **2017**, *2*, 1750017. [CrossRef]
29. Buhairi, M.A.; Foudzi, F.M.; Jamhari, F.I.; Sulong, A.B.; Radzuan, N.A.M.; Muhamad, N.; Mohamed, I.F.; Azman, A.H.; Harun, W.S.W.; Al-Furjan, M.S.H. Review on volumetric energy density: Influence on morphology and mechanical properties of Ti6Al4V manufactured via laser powder bed fusion. *Prog. Addit. Manuf.* **2023**, *8*, 265–283. [CrossRef]
30. Wu, H.; Ren, Y.; Ren, J.; Liang, L.; Li, R.; Fang, Q.; Cai, A.; Shan, Q.; Tian, Y.; Baker, I. Selective laser melted AlSi10Mg alloy under melting mode transition: Microstructure evolution, nanomechanical behaviors and tensile properties. *J. Alloys Compd.* **2021**, *873*, 159823. [CrossRef]
31. Liu, W.; Chen, C.; Shuai, S.; Zhao, R.; Liu, L.; Wang, X.; Hu, T.; Xuan, W.; Li, C.; Yu, J.; et al. Study of pore defect and mechanical properties in selective laser melted Ti6Al4V alloy based on X-ray computed tomography. *Mater. Sci. Eng. A* **2020**, *797*, 139981. [CrossRef]
32. Traxel, K.D.; Bandyopadhyay, A. Selective laser melting of Ti6Al4V-B4C-BN in situ reactive composites. *J. Mater. Res. Technol.* **2022**, *18*, 2654–2671. [CrossRef]

33. Wang, J.; Zhu, R.; Liu, Y.; Zhang, L. Understanding melt pool characteristics in laser powder bed fusion: An overview of single- and multi-track melt pools for process optimization. *Adv. Powder Mater.* **2023**, *2*, 100137. [CrossRef]
34. Kooi, B.J.; Pei, Y.T.; De Hosson, J.T.M. The evolution of microstructure in a laser clad TiB–Ti composite coating. *Acta Mater.* **2003**, *51*, 831–845. [CrossRef]
35. De Barros, M.I.; Rats, D.; Vandenbulcke, L.; Farges, G. Influence of internal diffusion barriers on carbon diffusion in pure titanium and Ti–6Al–4V during diamond deposition. *Diamond Relat. Mater.* **1999**, *8*, 1022–1032. [CrossRef]
36. Kværndrup, F.B.; Somers, M.A.J.; Christiansen, T.L. Extreme Expansion and Reversible Hydrogen Solubility in h.c.p. Titanium Stabilized by Colossal Interstitial Alloying. *Metall. Mater. Trans. A* **2021**, *52*, 4997–5003. [CrossRef]
37. Peter Rogl, H.B.; Duschanek, H. *Materials Science International Team. Calculated Isopleth from Ti to 'B4C': Datasheet from MSI Eureka in SpringerMaterials;* MSI, Materials Science International Services GmbH: Stuttgart, Germany, 2004; Available online: https://materials.springer.com/msi/phase-diagram/docs/sm_msi_r_10_011050_03_full_LnkDia11 (accessed on 19 June 2022).
38. Hasib, M.T.; Ostergaard, H.E.; Liu, Q.; Li, X.; Kruzic, J.J. Tensile and fatigue crack growth behavior of commercially pure titanium produced by laser powder bed fusion additive manufacturing. *Addit. Manuf.* **2021**, *45*, 102027. [CrossRef]
39. Liu, H.; Wang, H.; Ren, L.; Qiu, D.; Yang, K. Antibacterial copper-bearing titanium alloy prepared by laser powder bed fusion for superior mechanical performance. *J. Mater. Sci. Technol.* **2023**, *132*, 100–109. [CrossRef]
40. Issariyapat, A.; Kariya, S.; Shitara, K.; Umeda, J.; Kondoh, K. Solute-induced near-isotropic performance of laser powder bed fusion manufactured pure titanium. *Addit. Manuf.* **2022**, *56*, 102907. [CrossRef]
41. Attar, H.; Bermingham, M.J.; Ehtemam-Haghighi, S.; Dehghan-Manshadi, A.; Kent, D.; Dargusch, M.S. Evaluation of the mechanical and wear properties of titanium produced by three different additive manufacturing methods for biomedical application. *Mater. Sci. Eng. A* **2019**, *760*, 339–345. [CrossRef]
42. Okazaki, Y.; Ishino, A. Microstructures and Mechanical Properties of Laser-Sintered Commercially Pure Ti and Ti-6Al-4V Alloy for Dental Applications. *Materials* **2020**, *13*, 609. [CrossRef]
43. Wei, M.; Yu, H.; Song, Z.; Yin, Y.; Zhou, X.; Wang, H.; Ji, X.; Li, X.; Shi, P.; Zhang, W. Microstructural evolution, mechanical properties and wear behavior of in-situ TiC-reinforced Ti matrix composite coating by induction cladding. *Surf. Coat. Technol.* **2021**, *412*, 127048. [CrossRef]
44. Zheng, B.; Dong, F.; Yuan, X.; Zhang, Y.; Huang, H.; Zuo, X.; Luo, L.; Wang, L.; Su, Y.; Wang, X.; et al. Evaluation on tribological characteristics of (TiC+TiB)/Ti–6Al–4V composite in the range from 25 °C to 600 °C. *Wear* **2020**, *450–451*, 203256. [CrossRef]
45. Miller, R. Physical properties of liquid metals. In *Liquid Metals Handbook;* U.S. Government Publishing Office: Washington, DC, USA, 1952; Volume 38.
46. Nogi, K.; Ogino, K.; McLean, A.; Miller, W.A. The temperature coefficient of the surface tension of pure liquid metals. *Metall. Trans. B* **1986**, *17*, 163–170. [CrossRef]
47. Gäumann, M.; Bezençon, C.; Canalis, P.; Kurz, W. Single-crystal laser deposition of superalloys: Processing–microstructure maps. *Acta Mater.* **2001**, *49*, 1051–1062. [CrossRef]
48. Hu, Y.; Cong, W.; Wang, X.; Li, Y.; Ning, F.; Wang, H. Laser deposition-additive manufacturing of TiB-Ti composites with novel three-dimensional quasi-continuous network microstructure: Effects on strengthening and toughening. *Compos. Part B* **2018**, *133*, 91–100. [CrossRef]
49. Wang, J.; Guo, X.; Qin, J.; Zhang, D.; Lu, W. Microstructure and mechanical properties of investment casted titanium matrix composites with B4C additions. *Mater. Sci. Eng. A* **2015**, *628*, 366–373. [CrossRef]
50. Huang, L.; Kong, F.; Chen, Y.; Xiao, S. Microstructure and tensile properties of Ti–6Al–4V–0.1B alloys of direct rolling in the near β phase region. *Mater. Sci. Eng. A* **2013**, *560*, 140–147. [CrossRef]
51. Ren, Y.; Wu, H.; Liu, B.; Shan, Q.; Guo, S.; Jiao, Z.; Baker, I. A novel L12-strengthened AlCoCuFeNi high-entropy alloy with both high hardness and good corrosion resistance. *Mater. Lett.* **2023**, *331*, 133339. [CrossRef]
52. Scheu, C.; Stergar, E.; Schober, M.; Cha, L.; Clemens, H.; Bartels, A.; Schimansky, F.P.; Cerezo, A. High carbon solubility in a γ-TiAl-based Ti–45Al–5Nb–0.5C alloy and its effect on hardening. *Acta Mater.* **2009**, *57*, 1504–1511. [CrossRef]
53. Fan, Z.; Miodownik, A.P.; Chandrasekaran, L.; Ward-Close, M. The Young's moduli of in situ Ti/TiB composites obtained by rapid solidification processing. *J. Mater. Sci.* **1994**, *29*, 1127–1134. [CrossRef]
54. Liu, J.; Sun, Q.; Zhou, C.a.; Wang, X.; Li, H.; Guo, K.; Sun, J. Achieving Ti6Al4V alloys with both high strength and ductility via selective laser melting. *Mater. Sci. Eng. A* **2019**, *766*, 138319. [CrossRef]
55. Sarkar, S.; Kumar, C.S.; Nath, A.K. Effects of heat treatment and build orientations on the fatigue life of selective laser melted 15-5 PH stainless steel. *Mater. Sci. Eng. A* **2019**, *755*, 235–245. [CrossRef]
56. Yang, Y.; Ren, Y.; Tian, Y.; Li, K.; Zhang, W.; Shan, Q.; Tian, Y.; Huang, Q.; Wu, H. Microstructure and properties of FeCoCrNiMoSix high-entropy alloys fabricated by spark plasma sintering. *J. Alloys Compd.* **2021**, *884*, 161070. [CrossRef]
57. Deshpande, P.K.; Lin, R.Y. Wear resistance of WC particle reinforced copper matrix composites and the effect of porosity. *Mater. Sci. Eng. A* **2006**, *418*, 137–145. [CrossRef]

materials

MDPI

Article

The Effects of Co on the Microstructure and Mechanical Properties of Ni-Based Superalloys Prepared via Selective Laser Melting

Xiaoqiong Ouyang [1,2], Feng Liu [1,2], Lan Huang [1,2], Lin Ye [1,2], Heng Dong [1,2], Liming Tan [1,2,3,*], Li Wang [1,2,*], Xiaochao Jin [4] and Yong Liu [1,2]

1 State Key Laboratory of Powder Metallurgy, Central South University, Changsha 410083, China
2 Powder Metallurgy Research Institute, Central South University, Changsha 410083, China
3 Foshan (Southern China) Institute for New Materials, Foshan 528200, China
4 State Key Laboratory for Strength and Vibration of Mechanical Structures, Xi'an Jiaotong University, Xi'an 710049, China
* Correspondence: limingtan@csu.edu.cn (L.T.); li.wang@csu.edu.cn (L.W.)

Abstract: In this work, two Ni-based superalloys with 13 wt.% and 35 wt.% Co were prepared via selective laser melting (SLM), and the effects of Co on the microstructure and mechanical properties of the additively manufactured superalloys were investigated. As the Co fraction increased from 13 wt.% to 35 wt.%, the average grain size decreased from 25.69 μm to 17.57 μm, and the size of the nano-phases significantly increased from 80.54 nm to 230 nm. Moreover, the morphology of the γ' phase changed from that of a cuboid to a sphere, since Co decreased the γ/γ' lattice mismatch from 0.64% to 0.19%. At room temperature, the yield strength and ultimate tensile strength of the 13Co alloy reached 1379 MPa and 1487.34 MPa, and those of the 35Co alloy were reduced to 1231 MPa and 1350 MPa, while the elongation increased by 52%. The theoretical calculation indicated that the precipitation strengthening derived from the γ' precipitates made the greatest contribution to the strength.

Keywords: Ni-base superalloy; selective laser melting; cobalt; strength; ductility

1. Introduction

Ni-based superalloys have been broadly applied in aerospace industries due to their excellent high-temperature properties, such as their high strength and oxidation resistance [1,2]. In general, they are often used for hot-section components of engines, such as burners, turbine discs, turbine blades, etc. [3]. Significantly, the quality of a turbine engine is closely related to improvements in Ni-based superalloy performance. Traditionally, superalloy components were prepared by using casting, forging, and hot extrusion [4]. In contrast, these methods have disadvantages, such as complex preparation processes, low efficiency, and high costs [5], making it difficult for them to meet the requirements for forming integrated components with complex structures. The selective laser melting (SLM) technology can be used to directly shape powder particles, and the laser can be changed according to the geometry of the products. Hence, the preparation process for SLM is relatively simple, and it can be suitable for the preparation of complex structural parts. At the same time, it can ensure the dimensional accuracy of a product [6] and a good surface roughness [7].

With the application of SLM in Ni-based superalloys, researchers have tried to change the internal microstructures of the alloys by adjusting the elemental content to enhance the mechanical properties. Griffiths et al. [8] found that removing the Hf element could reduce the number of cracks and improve the yield strength of low-carbon CM247LC superalloys. In low-carbon IN738LC, the ultimate tensile strength was increased from 610 to 1113 MPa when the Zr content was reduced from 0.12 wt.% to 0.024 wt.% [9]. In addition, by increasing the Co fraction from 5 wt.% to 23 wt.%, Ni-Co-based superalloys exhibited a higher yield strength and ductility at 750 °C and 800 °C [10]. Murray et al. [11] obtained a crack-free Co-Ni-based superalloy with a high ultimate tensile strength of 1.1 GPa and elongation of over 13% at room temperature by increasing the Co content to 39 wt.%. Based on these studies, it is expected that the mechanical properties of additively manufactured Ni-based superalloys could be enhanced by adjusting the Co content.

Co can substitute Ni in the crystal lattice, and it is broadly deemed one of the main elements in γ'-strengthened Ni-based superalloys. Based on previous works, it was found that Co can cause γ'-$Ni_3(Al,Ti)$ to transform into γ'-$(Ni,Co)_3(Al,Ti)$ [12–14], reduce the stacking fault energy, and modify the lattice mismatch of γ/γ' to influence the morphology of γ' precipitates [15,16]. However, the effects of Co on other precipitates, such as carbides and the σ phase, have not yet been studied. Actually, except for γ' precipitates, factors originating from other precipitates, i.e., their morphology, size, distribution, and stability in superalloys, also have an important influence on the mechanical properties [17–19].

In this research, two superalloys with different Co contents were prepared via selective laser melting, and the γ' precipitates, carbides, and σ phase formed in the alloys were characterized. The effects of Co on the morphology and size of the precipitates were systematically analyzed, and the contributions to the strength from different strengthening mechanisms are discussed in detail.

2. Materials and Methods

2.1. Materials

γ'-strengthened Ni-based superalloys with different Co contents were the research object of the experiments. Table 1 shows the nominal chemical compositions of two pre-alloyed powders with Co contents of 13 wt.% and 35 wt.%. Both powders were produced by using gas atomization, and the distribution range of two particles' sizes was 10–63 μm, with D_{50} being around 34.3 μm, where D_{50} means that nearly half of the powder particles exceeded the particle size value that it represents. Figure 1a,b show the morphology of the powders. There were some satellite powders for 13Co, and the powder distribution was more uniform than that of 35Co.

Table 1. Nominal compositions of the two alloys (wt.%).

Alloy	Co	Cr	W	Mo	Ta	Ti	Al	C	B	Zr	Hf	Ni
13Co	13.00	11.96	4.03	4.02	4.02	4.03	2.99	0.06	0.04	0.03	0.14	Bal
35Co	35.00	12.30	4.59	3.67	4.00	3.02	2.43	0.06	0.04	0.04	0.14	Bal

2.2. SLM Process

The Farsoon 271M SLM device was used to prepare rectangular samples with dimensions of 28 × 12 × 14 mm^3. Argon was used as a protective atmosphere during the printing process. The following process parameters were used for preparation: a laser power of 200 W, a scanning speed of 1200 mm/s, a hatch space of 80 μm, and a layer thickness of 40 μm. Stainless steel material was used as the substrate and was preheated to 100 °C before printing. The strategy of bidirectional scanning with an interlayer rotation of 67°, as illustrated in Figure 1c, was chosen to reduce the residual stress between the layers [20].

2.3. Heat Treatment Process

The as-printed samples were heated at 1180 °C for super-solution treatment for 1 h and then set in the air to cool to room temperature. After that, aging treatments were conducted in two stages: first, the samples were heated at 650 °C for 24 h, followed by allowing them cool down in the air; second, the samples were deposited at 760 °C for 16 h, and then allowed to cool down in the air.

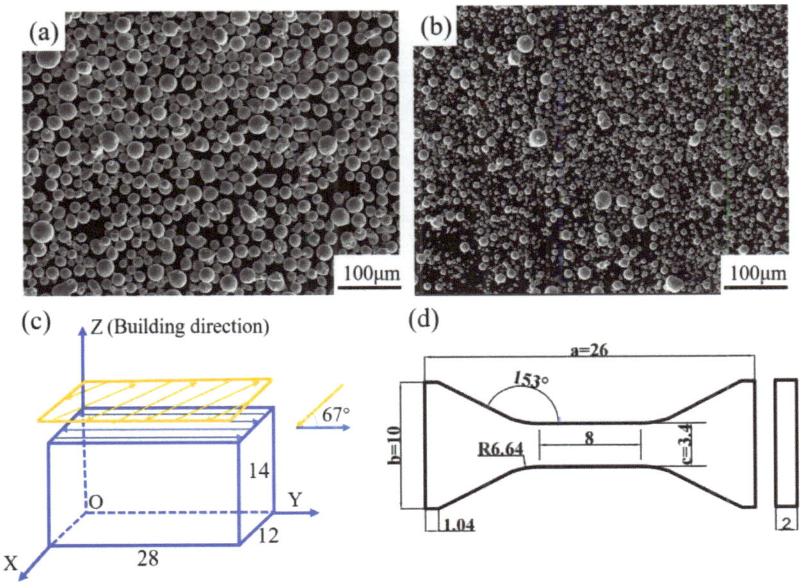

Figure 1. Secondary electron (SE) images of the powder morphology for (**a**) 13Co and (**b**) 35Co; the schematic images illustrate (**c**) the laser-scanning strategy and the size of a stretched specimen; (**d**) the geometry of the tensile specimen with a thickness of 2 mm.

2.4. Microstructural Characterization

Archimedes' principle was used to measure the relative density of the as-printed samples. Before microstructural characterization, SiC papers were used to sand the surfaces of the samples. Subsequently, all samples were polished to obtain mirror-like surfaces. All samples were etched with a Kalling reagent consisting of 50 g of $CuCl_2$, 100 mL of HCl, and 100 mL of methanol for observation with scanning electron microscopy (SEM, FEI Quanta 650 FEG, Tescan, Brno, Czech Republic). An electron backscattered diffraction (EBSD) investigation was undertaken at a voltage of 20 kV, a magnification of 500, and a step size of 2 μm. The data were processed by using the Channel 5 software. An X-ray diffraction (XRD) investigation was conducted on a Bruker D8 Advanced X-ray diffractometer(Bruker, Saarbrücken, Germany) with a scanning speed of 5°/min to measure the phase constitution and lattice parameters of the samples after heat treatment. The heat-treated samples were ground to a thickness of 50 μm and then subjected to ion thinning for transmission electron microscopy (TEM) observations. The TEM observations were conducted by using a Thermo Scientific™ Titan Spectra 300(Thermofisher, New York, NY, USA) equipped with a quadrant Super-X detector. The high-resolution scanning TEM (HRSTEM) and EDS results were collected and analyzed by using the Thermo Scientific™ Velox software.

2.5. Mechanical Properties

Tensile tests were conducted on the UTM5105 electronic universal testing machine. Tensile specimens were cut from rectangular samples by using the wire-cutting technique, and the profile of the tensile model is illustrated in Figure 1d. All tests were executed according to the ASTM E8 and E21 standards. The size of all tensile specimens in this work was 26 × 10 × 2 mm^3. The tensile tests were conducted at 25 °C and 850 °C with a strain rate of 10^{-3} s^{-1}. All tensile tests were repeated three times.

3. Results and Discussion

3.1. As-Printed Microstructure

The relative densities of the as-printed13Co and 35Co samples as measured with Archimedes' drainage method were both 99.6%, which indicated that the degrees of internal defects in 13Co and 35Co were the same. The backscattered electron (BSE) images of the XZ planes in the as-printed 13Co and 35Co alloys are illustrated in Figure 2a,b, respectively. Although the laser power, scanning speed, and hatch space were optimized to some extent, some cracks still appeared along the melt pools, and a few holes existed. Melt pools appeared in the XZ plane in the form of a "V" shape, and Figure 2c,d show that the melt pools in the XY plane were elliptical. The morphology of the melt pools was mainly related to the interaction between the laser and the powder [21] during the preparation of SLM. The absorption rate of the powder with the laser determined the degree of the laser's penetration of the powder and the depth of the melt pools, thereby affecting the morphology of the melt pools [22].

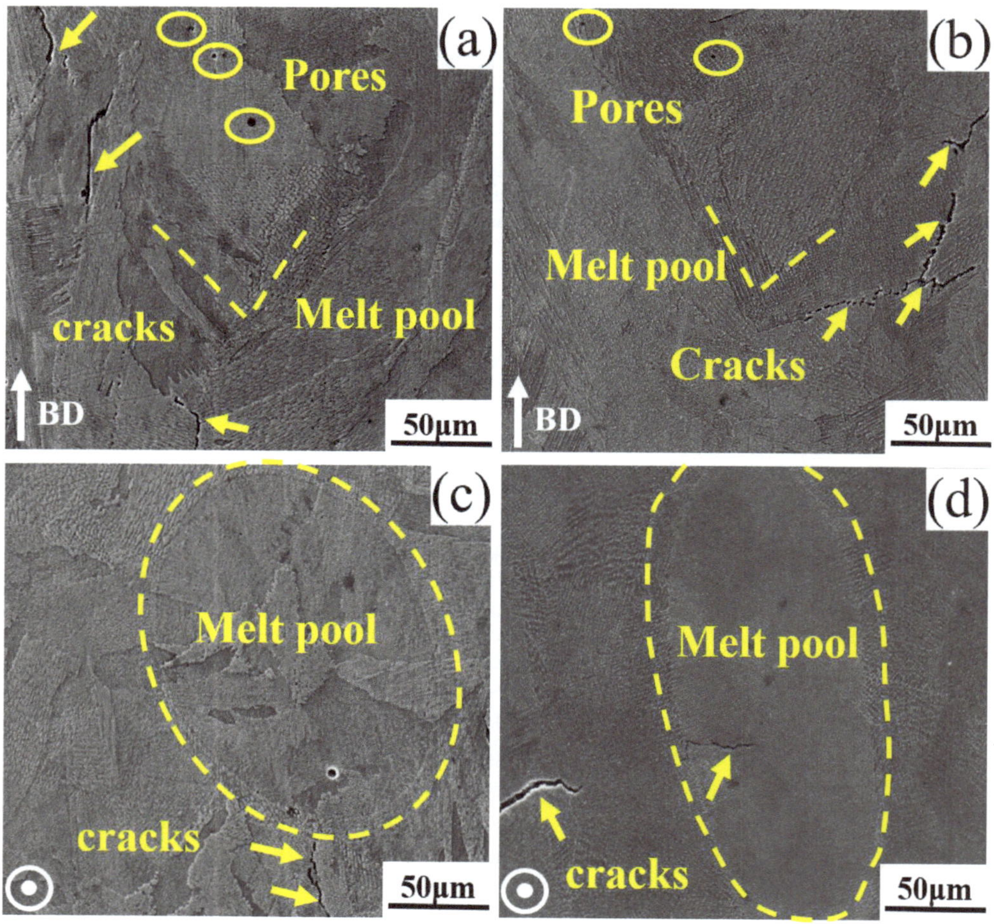

Figure 2. BSE images of as-printed microstructures in (**a**) the XZ plane in 13Co, (**b**) the XZ plane in 35Co, (**c**) the XY plane in 13Co, and (**d**) the XY plane in 35Co. The XZ plane was parallel to the building direction, and the XY plane was perpendicular to the building direction.

3.2. Grains

As shown in the inverse pole figures (IPFs) shown in Figure 3a,b, the grain size in the XY plane of the 13Co alloy ranged from 3 to 60 μm, and the average grain size was about 8 μm. The situation was similar in the 35Co alloy, where the range of the grain size in the XY plane was 2–63 μm, and the average grain size was about 8.96 μm. Thus, the difference in grain size between the two as-printed alloys was insignificant. Additionally, this displayed that the XY planes of both alloys presented the phenomenon of "small grains surrounding a large grain", which was mainly caused by the laser-scanning strategy with an interlayer rotation of 67°. It is worth mentioning that there were significantly more small grains around the large grains in the 35Co alloy compared to the 13Co alloy.

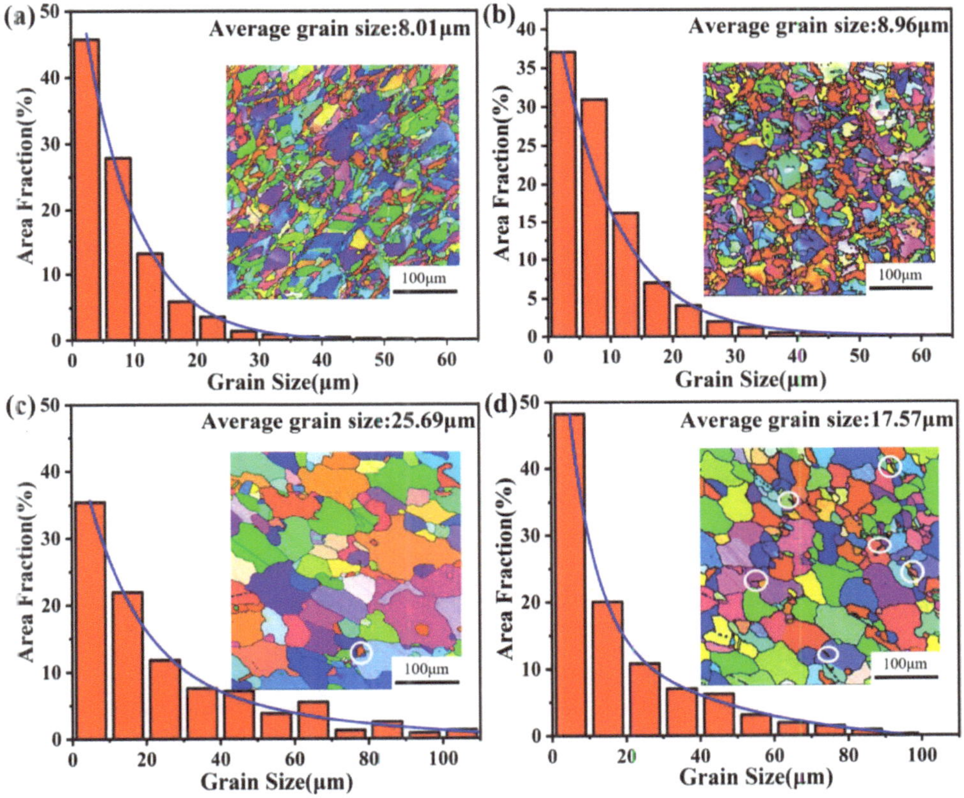

Figure 3. IPF maps obtained from the EBSD analysis in the XZ plane of the alloys and the corresponding grain size distribution diagram: (**a**) as-printed 13Co alloy; (**b**) as-printed 35Co alloy; (**c**) heat-treated 13Co alloy; (**d**) heat-treated 35Co alloy. The white circles represent twins.

After the super-solution treatment and aging treatments in two stages, the grains of the 13Co and 35Co alloys became coarse. However, the grains of the 35Co alloy were finer than those of the 13Co alloy, and the phenomenon of "small grains surrounding large grains" disappeared in the XY plane for both alloys, which can be observed in Figure 3c,d. Twins could be discerned, as indicated by the white circles in the images, and more twins were observed in 35Co than in 13Co, since the stacking fault energy of the γ matrix decreased with the increase in Co content [23]. After heat treatment, recrystallization occurred, and the fractions of the recrystallized grains for the 13Co and 35Co alloys were obtained by using the Recrystallized Fraction Component Function in the Channel 5 software, as displayed in Figure 4. It can be seen in Figure 4a,c that the deformation degree of 13Co before heat treatment was larger than that of 35Co. After heat treatment, in the 13Co alloy, the fraction of the recrystallized grains accounted for above 90%, as shown in Figure 4b, while it was only close to 50% in the 35Co alloy, as shown in Figure 4d. This demonstrated that the recrystallization of the 13Co alloy might have been close to completion, while the recrystallization process still took place in the 35Co alloy.

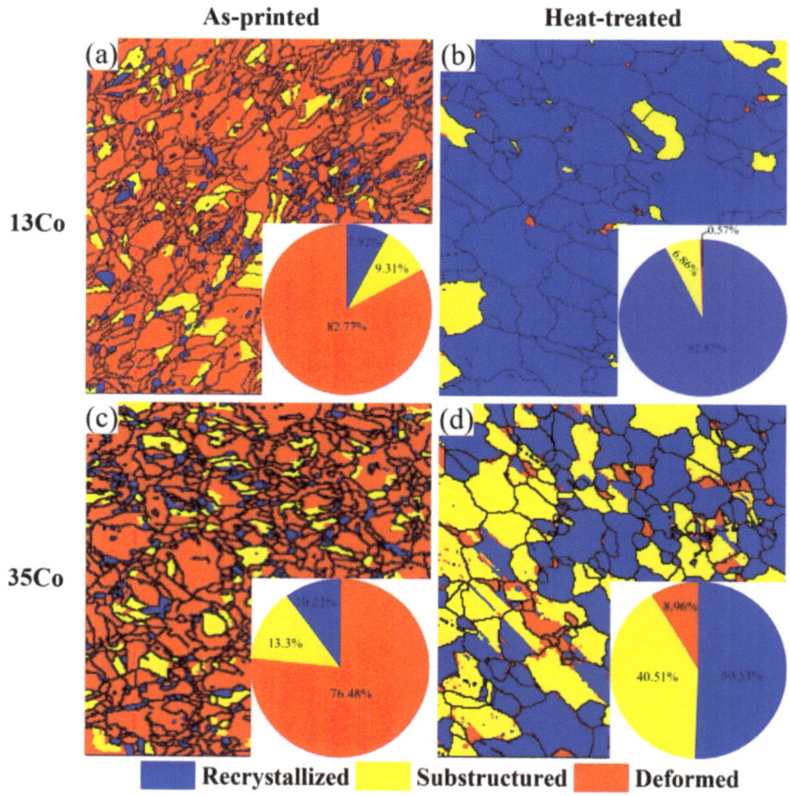

Figure 4. Recrystallization diagram of the XZ planes in (**a**) the as-printed 13Co, (**b**) heat-treated 13Co, (**c**) as-printed 35Co, and (**d**) heat-treated 35Co with recrystallized, substructured, and deformed grains.

According to previous studies [24,25], the greater the degree of deformation, the finer the grains will be after heat treatment. However, the grains of the 35Co alloy after heat treatment were finer than those of the 13Co alloy. Comparing Figure 3a,b, it can be seen that there were many small grains in the as-printed 35Co alloy, and more significant grain boundaries can provide more positions for nucleation. However, the internal deformation of 35Co was less than that of the as-printed 13Co alloy. Hence, the deformation energy could not provide the required driving force, resulting in finer grains after heat treatment.

3.3. Precipitate Distribution

3.3.1. XRD Analysis

The XRD spectra of the 13Co and 35Co alloys after heat treatment are shown in Figure 5, illustrating that the 13Co and 35Co alloys were substantially the same in terms of their phase configuration. The diffraction peaks in 35Co were all slightly shifted to the left, which was mainly caused by the increase in the Co content, since it increased the lattice parameters of γ and γ'. It should be noted that no reflections from other phases could be revealed in the XRD analysis, which was probably due to their small fractions. Thus, further electron microscopy investigations were conducted for the investigated alloys.

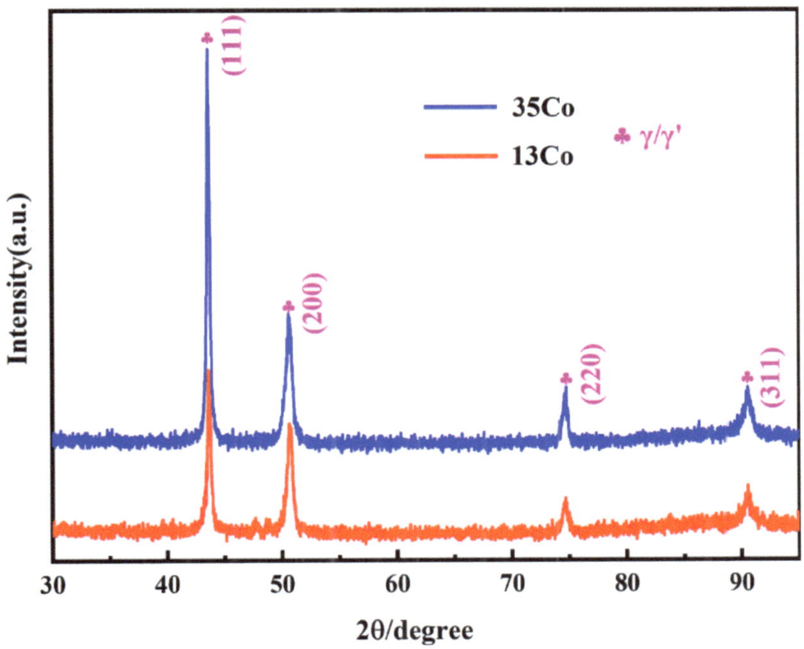

Figure 5. XRD spectra of the 13Co and 35Co alloys after heat treatment.

3.3.2. Formation Mechanisms and the Difference in the Sizes of MC Carbide and the σ Phase

Figure 6a–d clarify that SEM was used to further examine the microstructures of the 13Co and 35Co alloys after heat treatment in the backscattered electron mode; precipitates formed in both alloys, as shown with the white contrast. The white precipitates randomly existed in the samples. In addition, they were more prominent in the 35Co alloy than in the 13Co alloy. SEM images recorded in five randomly selected regions were used to evaluate the average size and volume fraction of these white precipitates, which were determined to be 81 nm and 1.3% in 13Co and 230 nm and 2.1% in 35Co, respectively.

The white precipitates in the two alloys were further characterized by using TEM, as shown in Figure 7. From the high-angle annular dark field-scanning transmission electron microscope (HAADF-STEM) micrograph and EDS maps of the 13Co alloy (Figure 7a), it can be inferred that the white precipitates were composed of two different phases, which was mainly because the elements Ta, Ti, and C were concentrated on one white precipitate and Mo and W were concentrated on the other white precipitate. A similar phenomenon appears in Figure 7b, so it can be tentatively judged that the types of the white precipitates were the same in 13Co and 35Co alloys.

Based on the EDS mapping, the white precipitates in 13Co and 35Co were probably the same. Figure 8 shows a further investigation of the two white precipitates in the 13Co alloy with high-resolution TEM to clarify their structures. The existence of the standard FCC-structured MC-phase carbide shown in Figure 8a was confirmed by the high-resolution image in Figure 8c and the associated fast Fourier transform (FFT) pattern along the axis of the $[0,\bar{1},1]$ zone. Figure 8b exhibits the formation of the tetragonally structured σ phase, which was confirmed by the high-resolution image and the associated FFT pattern along the axis of the [$\bar{1}$,0,0] zone shown in Figure 8d.

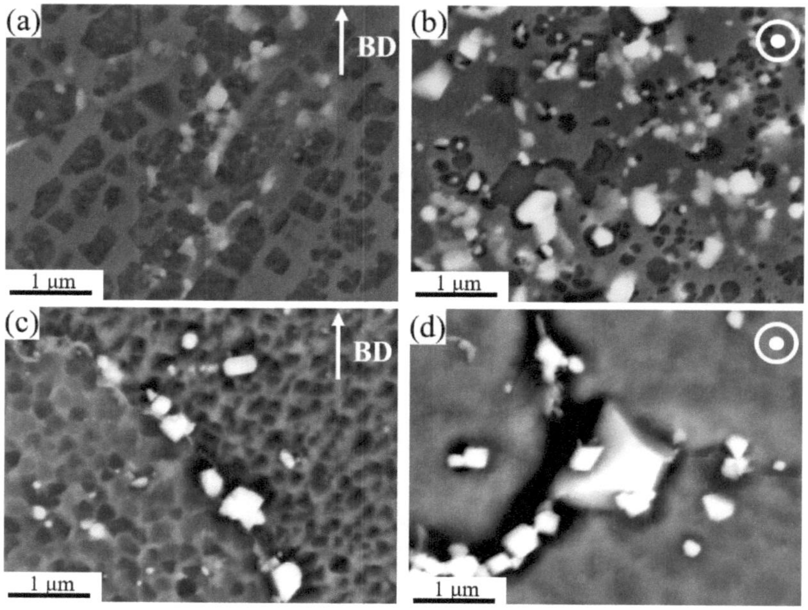

Figure 6. BSE images showing the microstructure after heat treatment of (**a**) the XZ plane of 13Co, (**b**) the XY plane of 13Co, (**c**) the XZ plane of 35Co, and (**d**) the XY plane of 35Co.

Through comprehensive investigations with SEM and TEM, the white precipitates appearing in the two alloys were identified as MC carbide and the σ phase. The formation of MC carbide could be attributed to the segregation of Ta, Ti, and C during the final solidification stage [26]. In Ni-based superalloys, Ta and Ti have stronger affinity with C than other elements do [27,28], which can prevent the combination of other elements with C, which would result in the formation of different kinds of carbides and the inhibition of the decomposition of MC carbides. It may be that the consumption of γ' precipitates relative to the Al and Ti elements was less significant, thereby promoting the formation and growth of MC carbides.

Mo and W are the main elements that formed the σ phase, as this is one of the TCP phases [29,30]. Sites with defects tend to favor the nucleation position of the σ phase, which is related to the low activation energy around them [31]. The ranking of defects in order of increasing activation energy is as follows: grain boundaries < twin boundaries < dislocations [31,32]. Therefore, the σ phase is difficult to nucleate and grow in recrystallized grains. Due to the mass of dislocations inside the non-recrystallized grains, many nucleation checkpoints can be provided for the σ phase [33]. The growth of the σ phase is mainly affected by atomic diffusion, which often needs grain boundaries and twin boundaries to provide paths. By comparing the recrystallization profiles of the 13Co and 35Co alloys, it could be seen that the number of non-recrystallized grains in 35Co after heat treatment was relatively high, so there were many nucleation checkpoints for the σ phase, and the number of grain boundaries and twin boundaries in 35Co was more significant, so it could provide favorable diffusion paths for the growth of the σ phase.

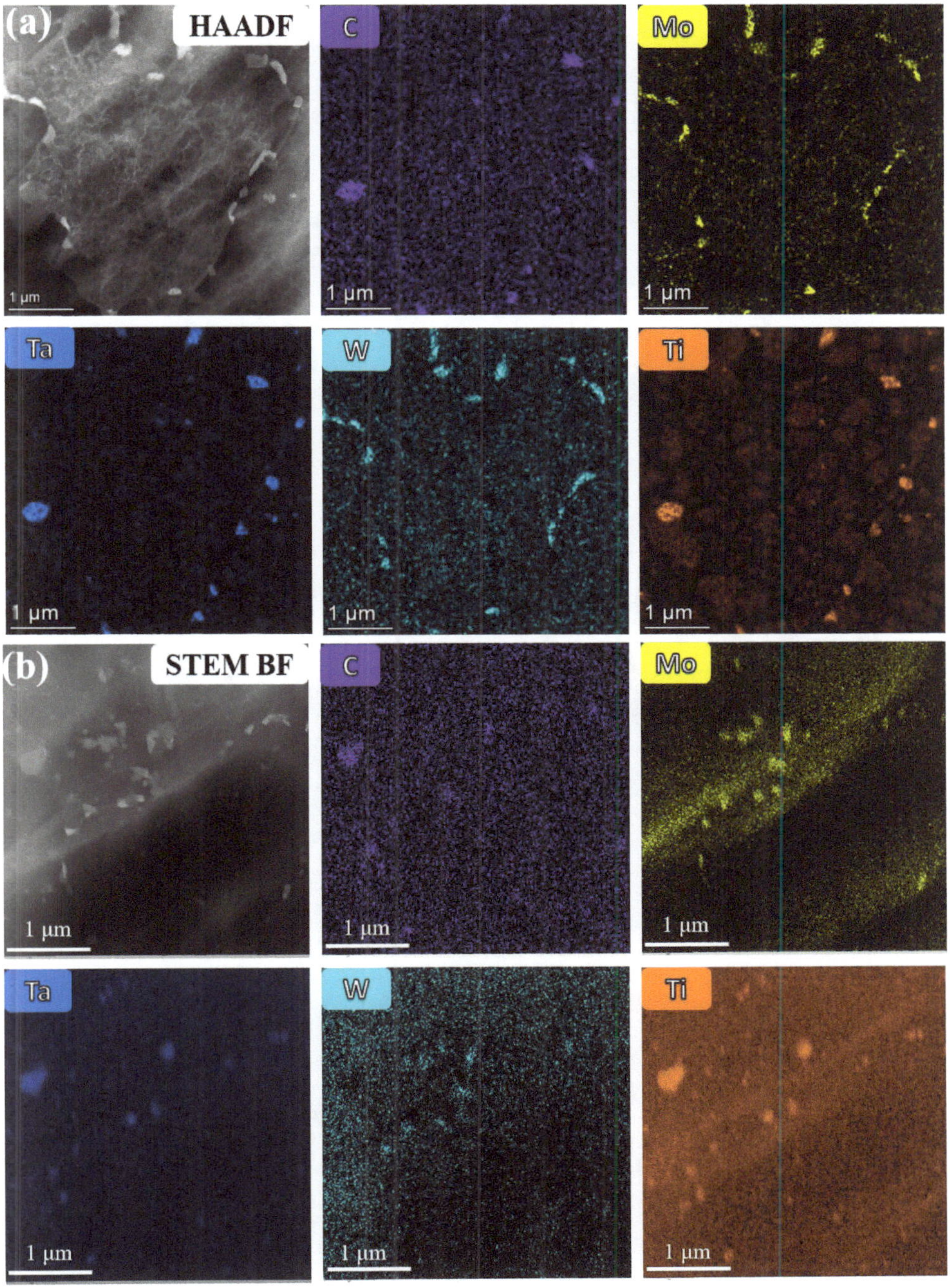

Figure 7. (**a**) HAADF-STEM image of the white precipitates and corresponding EDS mapping of 13Co. (**b**) Bright-field (BF) STEM image of the white precipitates and corresponding EDS mapping of 35Co.

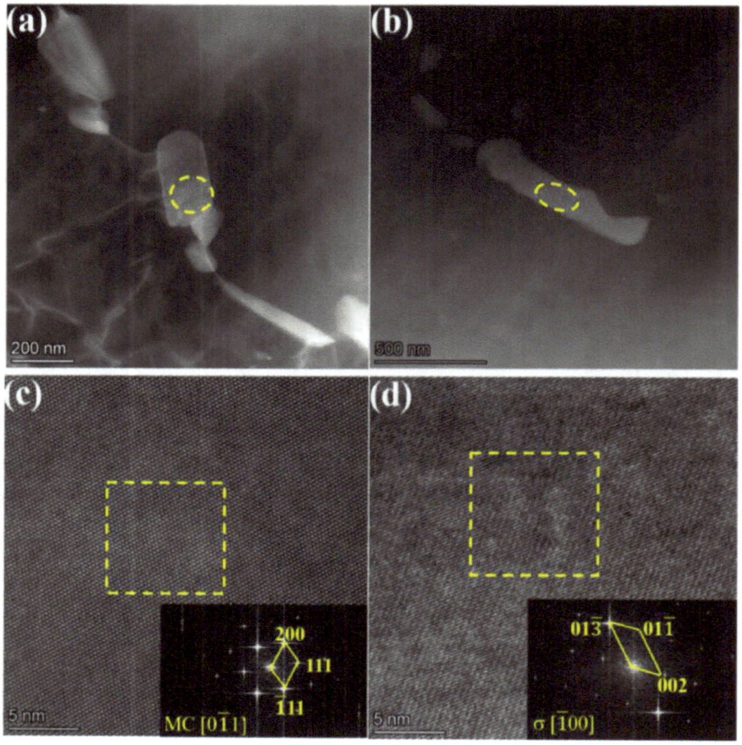

Figure 8. TEM investigation of the precipitates formed in the 13Co alloy. (**a**) HAADF-STEM image of MC carbides; (**b**) HAADF-STEM image of the σ phase; (**c**) HAADF-HRSTEM image and the corresponding FFT pattern of the MC carbide; (**d**) HAADF-HRSTEM image and the corresponding FFT pattern of the σ phase.

3.4. Transformation Mechanism of the γ' Morphology

Figure 9 shows HAADF-STEM images of the γ' precipitates and corresponding EDS maps in the 13Co and 35Co alloys. The γ' precipitates could not be quickly revealed in the HAADF-STEM images, but could be confirmed by the element maps. It could be seen that Ni, Al, Ti, and Ta were enriched in the γ' precipitates, and Co, Cr, and Mo were enriched in the γ matrix. By analyzing five SEM images that were recorded in randomly selected regions of both alloys by using Image J, the volume fractions of the γ' precipitates were determined to be about 48% and 38%. Figure 9a reveals that the γ' morphology was cubic in the 13Co alloy, and the average size was about 239 nm. However, the γ' morphology in the 35Co alloy was spherical, as shown in Figure 9b, and the average size was about 187 nm. It can be concluded that there were obvious differences in the morphologies of the γ' precipitates.

Some studies have shown that the morphology of γ' precipitates is mainly related to the γ/γ' lattice mismatch [34–36]. Due to the difference in the atomic radii of Co and Ni, the addition of Co could affect the lattice parameters of γ and γ', thereby affecting the lattice mismatch of γ/γ'. To obtain the influence on the lattice parameters of γ,γ' and the lattice mismatch of γ/γ', as shown in Figure 5, a Gaussian function was used to fit the $(200)_{\gamma/\gamma'}$ peaks in the XRD patterns, and Figure 10 exhibits the fitting results. The lattice mismatch of γ/γ' was then calculated with the following equations:

$$a = \frac{\lambda}{2\mathrm{Sin}\theta}\sqrt{h^2 + k^2 + l^2} \tag{1}$$

$$\delta = \frac{2(a_{\gamma'} - a_{\gamma})}{a_{\gamma'} + a_{\gamma}} \quad (2)$$

where a_{γ} and $a_{\gamma'}$ are the lattice parameters of the γ and γ' phases, respectively. The lattice parameters of γ/γ' in 13Co and 35Co were calculated to be 0.3596/0.3619 nm and 0.3597/0.3604 nm, respectively, and the lattice mismatches of γ/γ' were 0.64% and 0.19%, respectively. Although the lattice mismatch of γ/γ' in 13Co exceeded 0.5%, it still remained cubic, which is considered normal [37]. Therefore, the morphology of the shift of the γ' phase from cubic to spherical resulted from the increase in Co content.

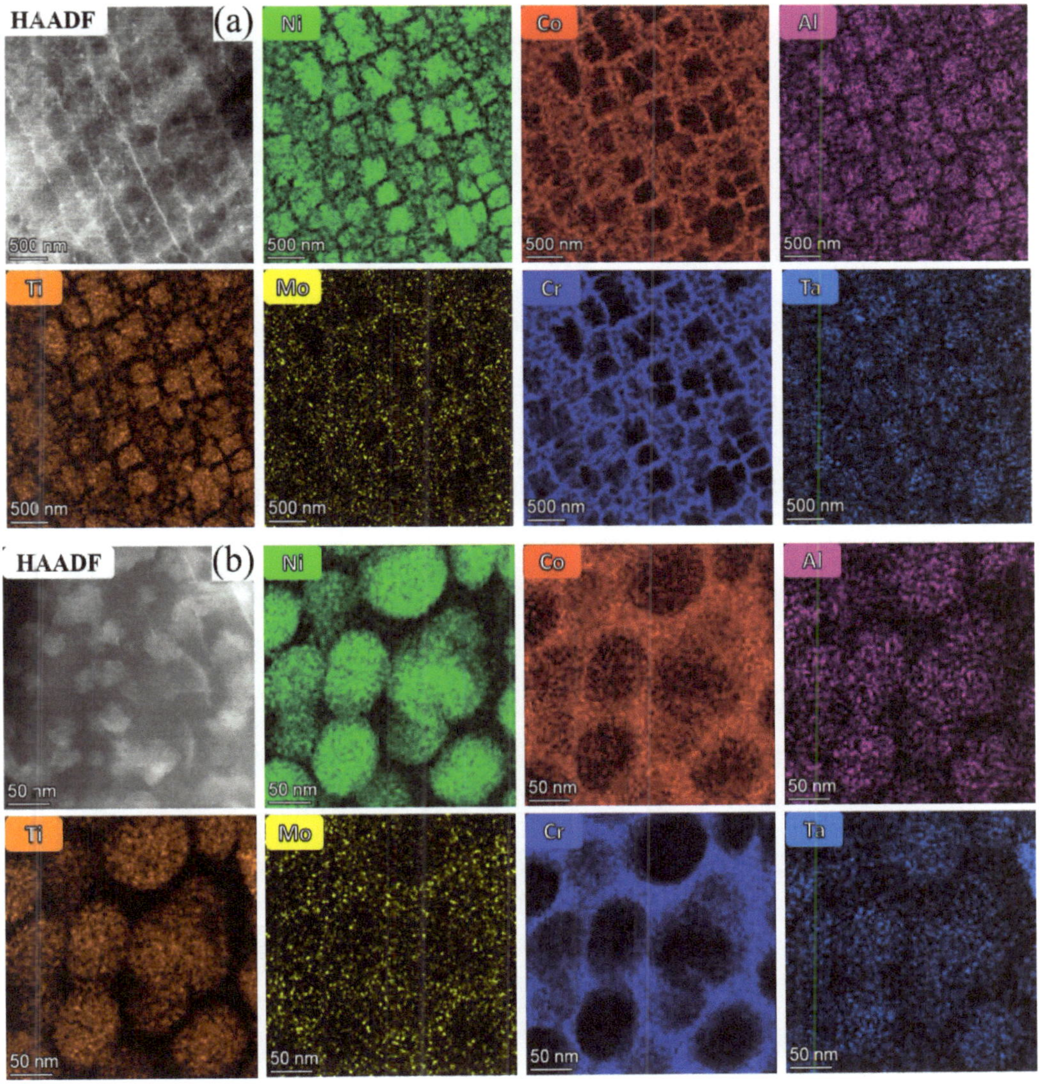

Figure 9. (**a**) HAADF-STEM images of the γ' precipitates and corresponding EDS mappings in 13Co. (**b**) HAADF-STEM images of the γ' precipitates and corresponding EDS mappings in 35Co.

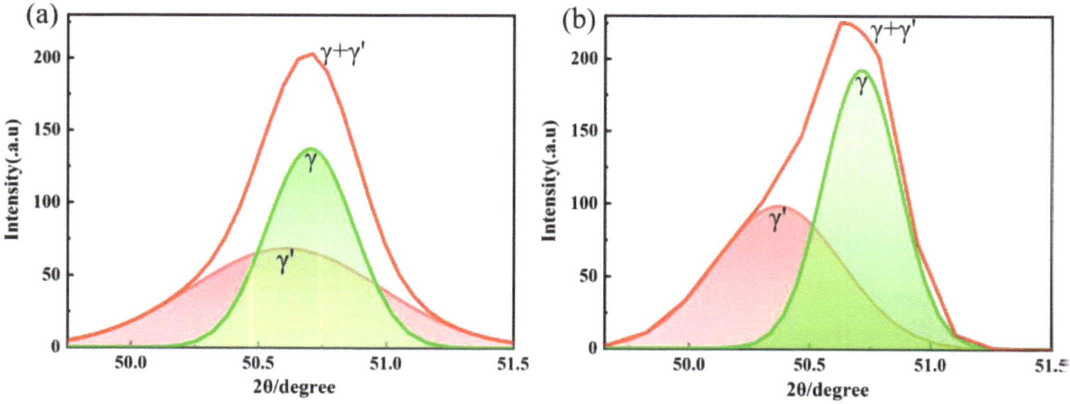

Figure 10. The (200) peaks of the XRD patterns for (**a**) 13Co and (**b**) 35Co with peak-fitted results with a Gaussian function.

3.5. Mechanical Properties

3.5.1. Fractured Morphology

The tensile properties of the heat-treated samples were tested at room temperature and 850 °C. The tensile results of all specimens are shown in Figure 11. At room temperature, the yield strength, ultimate tensile strength, and elongation were 1379 MPa, 1487 MPa, and 5.75% for the 13Co alloy and 1231 MPa, 1350 MPa, and 8.75% for the 35Co alloy, respectively. At 850 °C, the ultimate tensile strengths of the 13Co and 35Co alloys were around 329 MPa and 412 MPa, and their elongations were 2% and 2.25%, respectively. Compared to the 13Co alloy, the strength was increased by 25.5% and the ductility was slightly increased for the 35Co alloy. It is worth noting that the SLM 13Co and 35Co alloys after heat treatment showed better mechanical properties than those reported for Ni-based superalloys. Figure 12 shows the relationship between the yield strength and (Al + Ti) content. Based on the simple principle that alloys with (Al + Ti) content exceeding 6% are nonweldable [38] and a yield strength over 1200 MPa is deemed high, Figure 12 is divided into four regions.

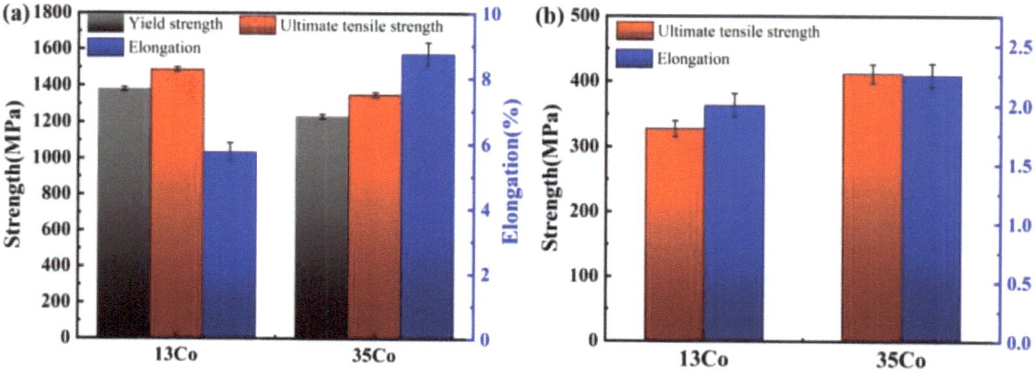

Figure 11. Mechanical properties of the 13Co and 35Co specimens after heat treatment at (**a**) room temperature and (**b**) 850 °C.

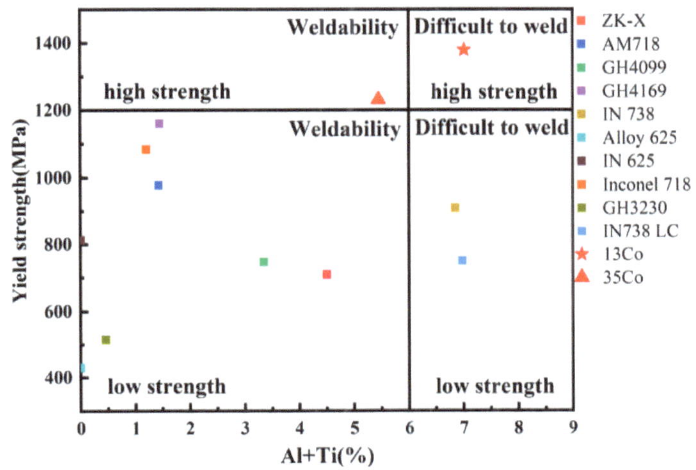

Figure 12. Comparison of the yield strength versus (Al + Ti)% for Ni-based superalloys in our work and other studies; all Ni-based superalloys were prepared via selective laser melting except for two superalloys, ZK-X and GH3230, which were prepared via laser melting deposition. Except for 13Co and 35Co alloys, the rest of the alloys are from the Refs. [39–48].

The fractured characteristics of the tensile specimens of 13Co and 35Co at room temperature are exhibited in Figure 13. Figure 13a,b demonstrate that there were some microcracks and pores on the fracture surface. Microcrack tips often acted as stress concentration points, causing accelerated crack propagation with extremely adverse effects on the tensile properties. In addition, a large number of dimples and intragranular tears were detected, as shown in Figure 13b,d. Judging from Figure 13c,f, the dimples of 35Co are smaller and deeper, indicating its better ductility than that of 13Co.

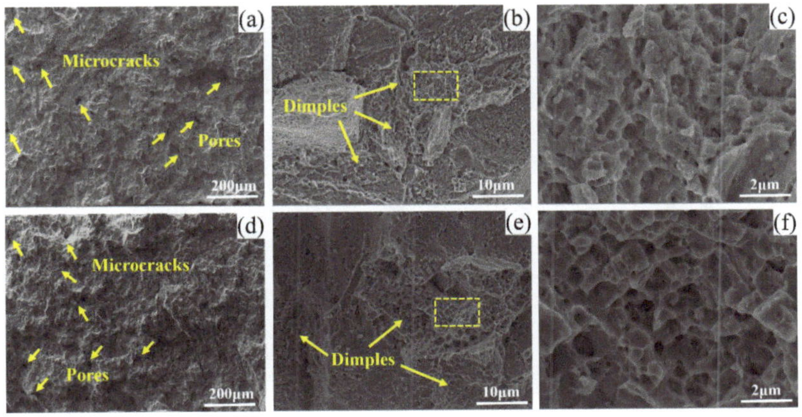

Figure 13. Room-temperature fractures of (**a–c**) 13Co and (**d–f**) 35Co.

3.5.2. Deformed Microstructures

In order to further understand the deformation of the structures during stretching, the side parts of the fractures were taken and characterized by using TEM. Figure 14a shows the stacking fault shearing of the γ' phase in 13Co; a large amount of dislocation was plugged at the grain boundary. Generally, when the dislocations encountered the coarse γ' particles, the shearing and dissociation of the dislocations usually caused stacking faults [10]. On the other hand, under stress, at the γ/γ' interface, the $\frac{1}{2}[110]$ dislocation dissociated into two Shockley dislocations, where 1/6[112] Shockley partials would shear the γ' precipitates, leading to the formation of stacking faults [49]. When a partial dislocation encountered a stacking fault, it could be hindered, resulting in T-shaped stacking faults. As shown in Figure 14b, there were full $\frac{1}{2}[110]$ dislocations in the matrix, and they often occurred as dislocation pairs at room temperature. Therefore, the ordered γ' phase was sheared by weakly coupled dislocation pairs or strongly coupled dislocation pairs. There was a large number of dislocation cells, as shown in Figure 14c, and the formation of dislocation cells was mainly related to the large dislocation density and the accumulation of dislocations. In addition, the phenomenon of nanoparticle-pinning dislocations was also found, as shown in Figure 14d. Fine and dispersed nanoparticles were strong barriers for dislocation glide [18]. Since the nanoparticles could hinder the movement of dislocations, dislocation entanglement occurred at the interface of incoherent nanoparticles and the γ matrix, resulting in local stress concentration and plasticity reduction.

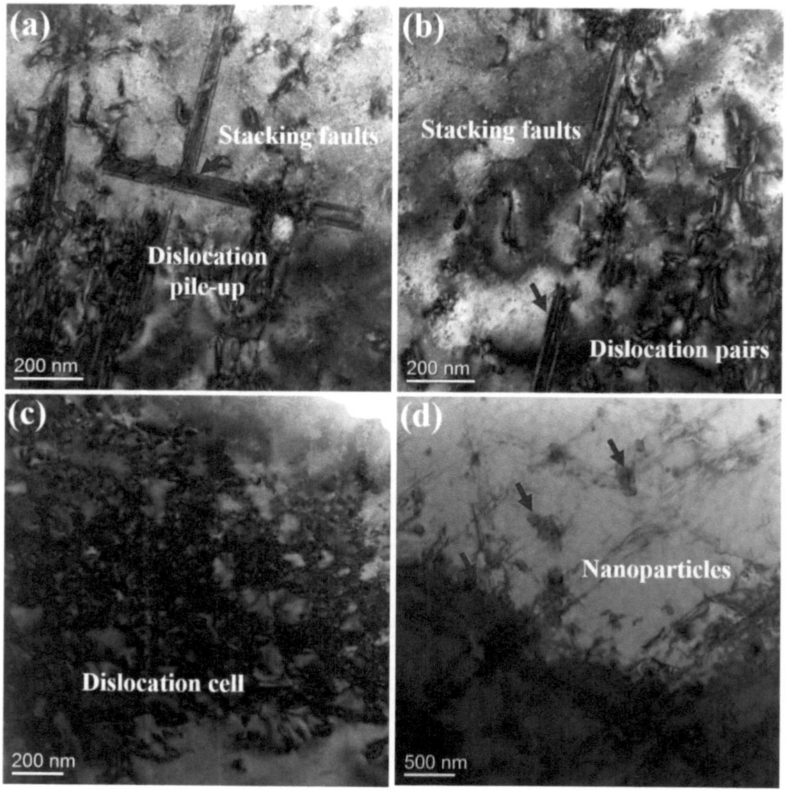

Figure 14. Deformed microstructure in the fractured specimen of 13Co: (**a**) stacking faults and dislocation pile-up; (**b**) stacking faults that sheared precipitates and interacted with dislocations; (**c**) dislocation cells; (**d**) nanoparticle-pinned dislocations.

More stacking faults and T-shaped configurations can be observed in Figure 15a,c. A large number of dislocation pairs that sheared γ' particles can be observed in Figure 15b; an APB-coupled dislocation pair that shears the γ' phase is a common deformation mechanism in Ni-based superalloys [50,51]. Deformed microtwins can also be observed in Figure 15d. Microtwins and stacking faults caused by dislocation dissociation play a key role in the strengthening process of alloys. The contributions of different strengthening mechanisms to the strength of the two alloys will be shown in the next section.

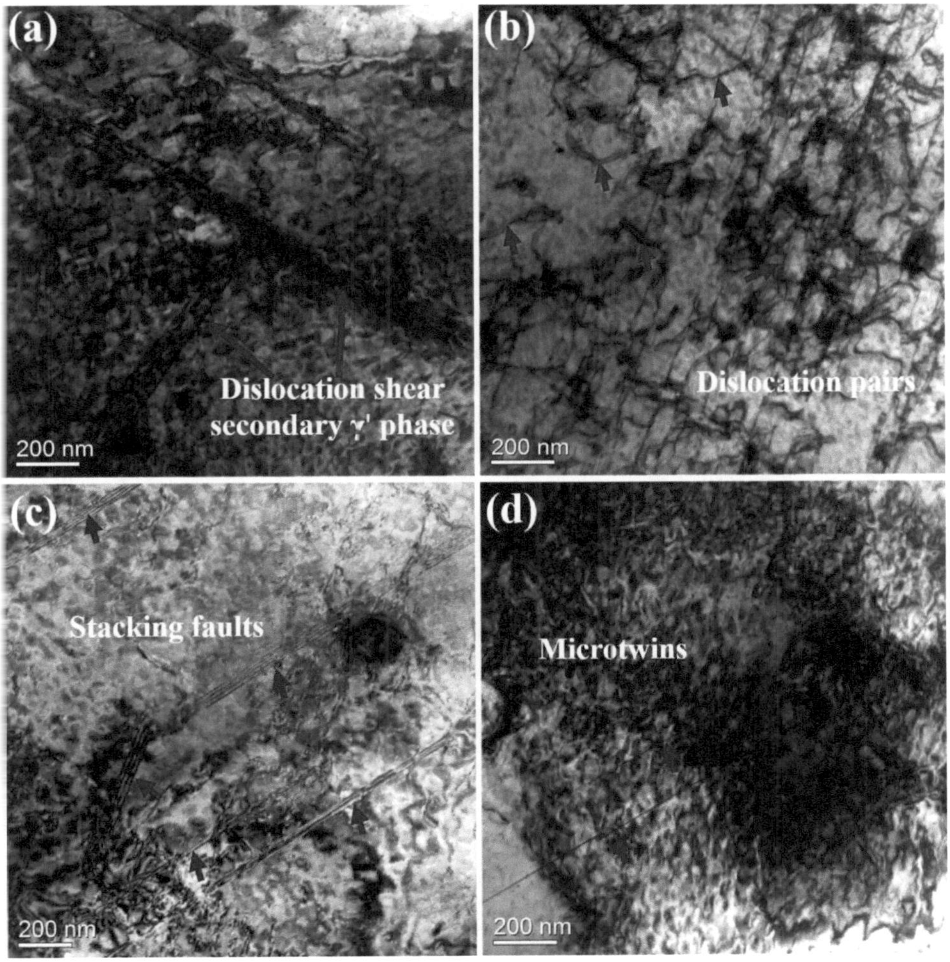

Figure 15. Deformed microstructure in the fractured specimen of 35Co: (**a**) dislocations that sheared secondary γ' precipitates; (**b**) dislocation pairs within γ' precipitates; (**c**) high-density stacking faults; (**d**) microtwins that cut through γ' precipitates.

3.5.3. Effect of Microstructure on Mechanical Properties

Based on the previous microstructural analysis of 13Co and 35Co, the differences were mainly reflected in the γ' phase, MC carbide, and σ phase. To explain the difference in mechanical properties between the two superalloys, a physical model was established to quantify the various effects on the yield strength of the alloys at room temperature. The yield strength $\sigma_{0.2}$ is the sum of the strength contributions of each mechanism, mainly including solid solution strengthening σ_{ss}, grain boundary strengthening σ_{Gb}, γ' phase strengthening σ_{p}, and nano-phase strengthening $\sigma_{\text{p-cut}}$.

$$\sigma_{0.2} = \sigma_{Ss} + \sigma_{Gb} + \sigma_{P} + \sigma_{p-cut} \tag{3}$$

Solid Solution Strengthening

For γ'-precipitation-strengthened Ni-based superalloys, L. Gypen [52] et al. proposed an expression for calculating σ_{SS}:

$$\sigma_{Ss} = (1 - f_{\gamma'})\left[\sum\left(\beta_i x_i^{1/2}\right)^2\right]^{1/2} \quad (4)$$

where $f_{\gamma'}$ represents the volume fraction of the γ' phase. x_i is the atomic percentage of the i-th element represented in the γ matrix, and β_i represents the solid solution strengthening coefficient of the i-th element, which is related to the elemental atomic radius and modulus. The β_i coefficient was determined with reference to [53].

Grain Boundary Strengthening

The grain boundary strengthening of Ni-based superalloys follows the Hall–Petch relationship and can be expressed as:

$$\sigma_{Gb} = K_{Hp} d^{-1/2} \quad (5)$$

where K_{HP} is an experimental constant related to the material properties and d is the average grain size. For superalloys, the value of K_{HP} ranges from 710 to 750 MPa·μm$^{-1/2}$ [50], and in this study, the K_{HP} value of 710 was used [54].

3.5.4. Precipitation Strengthening

Precipitation strengthening is the most essential strengthening mechanism in Ni-based superalloys. The critical resolved shear stress (CRSS), which is defined as a driving force that can make dislocation pairs pass through the γ' phase, is proportional to the precipitation intensity [54]. The precipitation strengthening can be divided into weak pair coupling, strong pair coupling, and the Orowan ring, which is based on the relationship between the size of the precipitates and the distances of the pairs of dislocations. According to Reed's theory [55], the shear stress of weak pair coupling, strong pair coupling, and pair coupling can be described as

$$\tau_{weak} = \frac{\gamma_{APB}}{2b}\left[\left(\frac{6\gamma_{APB} f r}{2\pi T}\right)^{1/2} - f\right] \quad (6)$$

$$\tau_{strong} = \sqrt{\frac{3}{2}}\left(\frac{Gb}{r}\right)\frac{f^{1/2}}{\pi^{2/3}}\left(\frac{2\pi\gamma_{APB}}{Gb^2} - 1\right)^{1/2} \quad (7)$$

where b and G represent the Burgers vector and shear modulus, respectively; this study used 0.254 nm [56] and 80 GPa [54,57,58]. T is the line tension of dislocation and is numerically equal to $(Gb^2)/2$. R and f represent the average size and volume fraction of the γ' phase. γ_{APB} is the APB energy, which represents the threshold required for dislocations to shear through the precipitate. Furthermore, γ_{APB} can be expressed as follows [59]:

$$\gamma_{APB} = \gamma_{APB}^0 + \sum_{i=1}^{n} k_i c_i \quad (8)$$

where c_i is the atomic fraction of solute i in the γ' phase, n is the number of solute elements in the γ' phase, and γ_{APB}^0 = 195 mJ/m^2 [60]. Here, the coefficient of APB energy change, k_i, is determined by density functional theory (DFT) with reference to [59].

When the size of the precipitates exceeds the critical radius of γ', the dislocation can bypass the γ' phase to form an Orowan ring, and the CRSS τ_{orowan} of the Orowan ring is described as follows [61]:

$$\tau_{orowan} = \frac{3Gb}{2L} \quad (9)$$

$$L = \left(\frac{2\pi}{3f}\right)^{1/2} r \tag{10}$$

$$\sigma_P = M\left(\tau_{weak}^n + \tau_{strong}^n + \tau_{orowan}^n\right) \tag{11}$$

where M is the Taylor factor, which was taken as 3.06 [57], and n is a fitting coefficient, which was taken as 5/6 in this work.

Nanoparticle Strengthening

According to the R-B model [62], precipitates will interact with a dislocation, thus hindering the movement of the dislocation and resulting in an increase in intensity [63]. Assuming that dislocations can be accordingly strengthened through weak particles, the expression is as follows [62]:

$$\sigma_{p-cut} = M\frac{Gb}{L}\left[1 - \left(\frac{E_p}{E_m}\right)^2\right]^{3/4} \tag{12}$$

where E_p represents the dislocation line energies in the nanoparticles and E_m represents the dislocation line energies in the γ matrix. L is the average distance between nanoparticles in the glide plane. The E_p/E_m ratio is 0.62 [64].

In the 13Co alloy, the strength contribution of each part was σ_{ss} = 220.85 MPa, σ_{Gb} = 140.08 MPa, σ_p = 961.69 MPa, and σ_{p-cut} = 41.64 MPa; in the 35Co alloy, the strength contribution of each part was established to be σ_{Ss} = 259.88 MPa, σ_{Gb} = 169.38 MPa, σ_p = 954.61 MPa, and σ_{p-cut} = 18.65 MPa. To sum up, the results of the calculated yield strengths of the 13Co and 35Co alloys were in agreement with the experimental results. In the 13Co and 35Co alloys, precipitation strengthening contributed the most, followed by solid solution strengthening. This indicated that the size and volume fraction of γ' precipitates had the greatest influence on the yield strength of the 13Co and 35Co alloys.

4. Conclusions

In this study, 13Co and 35Co alloys were successfully fabricated via SLM. The effects of Co on the microstructure and properties of the additively manufactured superalloys were systematically investigated, and the main conclusions can be summarized as follows.

(1) After heat treatment, the grain size of the 13Co alloy was coarser than that of the 35Co alloy, since the larger strain formed in the as-printed 13Co alloy could provide a driving force for grain growth.

(2) MC carbides and the σ phase were observed in both 13Co and 35Co, and their size increased from 80.54 nm to 230 nm with the increase in Co content. The increase in Co reduced the consumption of γ' relative to the Al and Ti content, thereby promoting the formation and growth of MC carbides. The size difference of the σ phase was mainly related to the large number of dislocations in the non-recrystallized grains after heat treatment.

(3) As the Co content increased from 13 wt.% to 35 wt.%, the lattice mismatch of γ/γ' decreased from 0.64% to 0.19%, and the morphology of the γ' precipitates shifted from cubic to spherical.

(4) At room temperature, the yield strength of 13Co was 10.7% higher than that of 35Co, but the ductility is relatively lower, since the 13Co alloy had more internal cracks and the dislocation entanglement and local stress at the interface of incoherent nanoparticles and the γ matrix led to a decrease in the elongation.

(5) The strengthening mechanism of the 13Co and 35Co alloys was dominated by precipitation strengthening, which contributed about 70% to $\sigma_{0.2}$, followed by solid solution strengthening.

Author Contributions: X.O.: writing—original draft, conceptualization, investigation. F.L.: writing—review and editing, supervision. L.H.: resources, supervision. L.Y.: data analysis, investigation. H.D.: investigation. X.J.: investigation. L.T.: resources, language polishing, and supervision. L.W.: resources, supervision. Y.L.: resources, supervision. All authors have read and agreed to the published version of the manuscript.

Funding: This research was funded by National Natural Science Foundation of China (52074366 and 52001334); National Science and Technology Major Project (J2019-IV-0003-0070); Natural Science Foundation of Hunan Province of China (2021JJ40757); Science and Technology Innovation Program of Hunan Province (2021RC3131).

Institutional Review Board Statement: Not applicable.

Informed Consent Statement: Not applicable.

Data Availability Statement: Not applicable.

Acknowledgments: This work was supported by the National Science and Technology Major Project (J2019-IV-0003-0070), the National Natural Science Foundation of China (52074366, 52001334), the Natural Science Foundation of Hunan Province of China (2021JJ40757), the Science and Technology Innovation Program of Hunan Province (2021RC3131), and a project supported by the State Key Laboratory of Powder Metallurgy, Central South University, Changsha, China.

Conflicts of Interest: The authors declare no conflict of interest.

References

1. Lee, C.M.; Han, Y.-S.; Jeong, J.S.; Jung, W.S.; Song, S.H.; Kim, J.-S. Effect of pre-strain on oxidation behaviour of nickel-based superalloys in air at 700 °C. *Corros. Sci.* **2022**, *200*, 110229–110241. [CrossRef]
2. Gao, Y.; Stölken, J.; Kumar, M.; Ritchie, R. High-cycle fatigue of nickel-base superalloy René 104 (ME3): Interaction of microstructurally small cracks with grain boundaries of known character. *Acta Mater.* **2007**, *55*, 3155–3167. [CrossRef]
3. Darolia, R. Development of strong, oxidation and corrosion resistant nickel-based superalloys: Critical review of challenges, progress and prospects. *Int. Mater. Rev.* **2018**, *64*, 355–380. [CrossRef]
4. Mazumder, Z. Laser Direct Metal Deposition Technology and Microstructure and Composition Segregation of Inconel 718 Superalloy. *J. Iron Steel Res. Int.* **2011**, *18*, 73–78.
5. Wang, R.; Jiang, K.; Jing, F.; Hu, D. Thermomechanical fatigue failure investigation on a single crystal nickel superalloy turbine blade. *Eng. Fail. Anal.* **2016**, *66*, 284–295. [CrossRef]
6. Khorasani, M.; Gibson, I.; Ghasemi, A.H.; Hadavi, E.; Rolfe, B. Laser subtractive and laser powder bed fusion of metals: Review of process and production features. *Rapid Prototyp. J.* 2023; *ahead-of-print*. [CrossRef]
7. Santosh, S.; Thomas, J.K.; Pavithran, M.; Nithyanandh, G.; Ashwath, J. An experimental analysis on the influence of CO_2 laser machining parameters on a copper-based shape memory alloy. *Opt. Laser Technol.* **2022**, *153*, 108210–108225. [CrossRef]
8. Griffiths, S.; Ghasemi-Tabasi, H.; De Luca, A.; Pado, J.; Joglekar, S.S.; Jhabvala, J.; Logé, R.E.; Leinenbach, C. Influence of Hf on the heat treatment response of additively manufactured Ni-base superalloy CM247LC. *Mater. Charact.* **2021**, *171*, 110815–110828. [CrossRef]
9. Hu, Y.; Yang, X.-K.; Kang, W.-J.; Ding, Y.-T.; Xu, J.-Y.; Zhang, H.-Y. Effect of Zr content on crack formation and mechanical properties of IN738LC processed by selective laser melting. *Trans. Nonferrous Met. Soc. China* **2021**, *31*, 1350–1362. [CrossRef]
10. Tang, L.; Liang, J.; Cui, C.; Li, J.; Zhou, Y.; Sun, X.; Ding, Y. Influence of Co content on the microstructures and mechanical properties of a Ni–Co base superalloy made by specific additive manufacturing process. *Mater. Sci. Eng. A* **2020**, *786*, 139438. [CrossRef]
11. Murray, S.P.; Pusch, K.M.; Polonsky, A.T.; Torbet, C.J.; Seward, G.G.E.; Zhou, N.; Forsik, S.A.J.; Nandwana, P.; Kirka, M.M.; Dehoff, R.R.; et al. A defect-resistant Co–Ni superalloy for 3D printing. *Nat. Commun.* **2020**, *11*, 4975–4986. [CrossRef]
12. Xu, L.; Cui, C.; Sun, X. The effects of Co and Ti additions on microstructures and compressive strength of Udimet710. *Mater. Sci. Eng. A* **2011**, *528*, 7851–7856. [CrossRef]
13. Oni, A.; Broderick, S.; Rajan, K.; LeBeau, J. Atom site preference and γ'/γ mismatch strain in NiAlCoTi superalloys. *Intermetallics* **2016**, *73*, 72–78. [CrossRef]
14. Llewelyn, S.; Christofidou, K.; Araullo-Peters, V.; Jones, N.; Hardy, M.; Marquis, E.; Stone, H. The effect of Ni:Co ratio on the elemental phase partitioning in γ-γ' Ni-Co-Al-Ti-Cr alloys. *Acta Mater.* **2017**, *131*, 296–304. [CrossRef]
15. Feng, Q.; Rowland, L.; Pollock, T. New Phases in a Multicomponent High Ruthenium Single Crystal Superalloy. *Mater. Sci. Forum* **2005**, *475–479*, 655–660. [CrossRef]
16. Li, Y.; Ye, X.; Li, J.; Zhang, Y.; Koizumi, Y.; Chiba, A. Influence of cobalt addition on microstructure and hot workability of IN713C superalloy. *Mater. Des.* **2017**, *122*, 340–346. [CrossRef]
17. Wen, P.; Wang, C.; Zhou, P.; Yang, M.; Wang, L.; Wang, B.; Wang, H. Effect of ultrasonic treatment on the microstructure and mechanical properties of a new ni-base superalloy 4716MA0. *J. Alloys Compd.* **2022**, *922*, 166278–166290. [CrossRef]

18. Chen, Z.; Li, Y.; Zhang, S.; Wei, P.; Dang, X.; Jiang, X.; Lu, B. Hierarchical architecture and mechanical behavior of K418 Ni-based superalloys manufactured by laser powder bed fusion. *Mater. Sci. Eng. A* **2022**, *851*, 143630–143640. [CrossRef]
19. Antonov, S.; Rozman, K.A.; Jablonski, P.D.; Detrois, M. Implications of carbon content for the processing, stability, and mechanical properties of cast and wrought Ni-based superalloy Nimonic 105. *Mater. Sci. Eng. A* **2022**, *857*, 144049–144065. [CrossRef]
20. Nadammal, N.; Mishurova, T.; Fritsch, T.; Serrano-Munoz, I.; Kromm, A.; Haberland, C.; Portella, P.D.; Bruno, G. Critical role of scan strategies on the development of microstructure, texture, and residual stresses during laser powder bed fusion additive manufacturing. *Addit. Manuf.* **2020**, *38*, 101792. [CrossRef]
21. Balasubramaniyan, C.; Rajkumar, K.; Santosh, S. Fiber Laser Cutting of Cu–Zr Added Quaternary NiTi Shape Memory Alloy: Experimental Investigation and Optimization. *Arab. J. Sci. Eng.* **2022**, *48*, 3665–3679. [CrossRef]
22. Khorasani, M.; Ghasemi, A.; Leary, M.; Sharabian, E.; Cordova, L.; Gibson, I.; Downing, D.; Bateman, S.; Brandt, M.; Rolfe, B. The effect of absorption ratio on meltpool features in laser-based powder bed fusion of IN718. *Opt. Laser Technol.* **2022**, *153*, 108263–108277. [CrossRef]
23. Yuan, Y.; Gu, Y.; Cui, C.; Osada, T.; Zhong, Z.; Tetsui, T.; Yokokawa, T.; Harada, H. Influence of Co content on stacking fault energy in Ni–Co base disk superalloys. *J. Mater. Res.* **2011**, *26*, 2833–2837. [CrossRef]
24. Adegoke, O.; Polisetti, S.R.; Xu, J.; Andersson, J.; Brodin, H.; Pederson, R.; Harlin, P. Influence of laser powder bed fusion process parameters on the microstructure of solution heat-treated nickel-based superalloy Alloy 247LC. *Mater. Charact.* **2021**, *183*, 111612. [CrossRef]
25. Engeli, R. Selective laser melting & heat treatment of γ' strengthened Ni-base superalloys for high temperature applications. Doctoral Dissertation, ETH Zurich Research Collection, ETH Zurich, Zürich, Switzerland, 2017.
26. Cao, S.; Yang, Y.; Chen, B.; Liu, K.; Ma, Y.; Ding, L.; Shi, J. Influence of yttrium on purification and carbide precipitation of superalloy K4169. *J. Mater. Sci. Technol.* **2021**, *86*, 260–270. [CrossRef]
27. Gao, S.; Song, Z.-F.; He, B.; Zhou, L.-Z.; Hou, J.-S. Effect of Ta addition on solidification microstructure and element segregation of IN617B nickel-base superalloy. *Trans. Nonferrous Met. Soc. China* **2022**, *32*, 559–568. [CrossRef]
28. Bai, J.; Zhang, H.; Liu, J.; Sun, Q.; Liu, C.; Zhang, Y. Investigation of room temperature strengthening mechanism on PM Ni-base superalloys with tantalum addition. *Mater. Charact.* **2022**, *191*, 112089–112100. [CrossRef]
29. Long, F.; Yoo, Y.; Jo, C.; Seo, S.; Song, Y.; Jin, T.; Hu, Z. Formation of η and σ phase in three polycrystalline superalloys and their impact on tensile properties. *Mater. Sci. Eng. A* **2009**, *527*, 361–369. [CrossRef]
30. Antonov, S.; Huo, J.; Feng, Q.; Isheim, D.; Seidman, D.; Helmink, R.; Sun, E.; Tin, S. σ and η Phase formation in advanced polycrystalline Ni-base superalloys. *Mater. Sci. Eng. A* **2017**, *687*, 232–240. [CrossRef]
31. Zhang, L.; Zhang, L.; Wang, H.; Li, J.; Man, J.; Xu, Z.; Yu, J.; Wan, G.; Wang, W.; Wu, B. Evolution of the microstructure and mechanical properties of an sigma-hardened high-entropy alloy at different annealing temperatures. *Mater. Sci. Eng. A* **2021**, *831*, 142140. [CrossRef]
32. Cahn, J.W. The kinetics of grain boundary nucleated reactions. *Acta Met.* **1956**, *4*, 449–459. [CrossRef]
33. Zhang, L.; Huo, X.; Wang, A.; Du, X.; Zhang, L.; Li, W.; Zou, N.; Wan, G.; Duan, G.; Wu, B. A ductile high entropy alloy strengthened by nano sigma phase. *Intermetallics* **2020**, *122*, 106813. [CrossRef]
34. Xu, Y.; Lv, X.; Ma, T.; Li, H.; Wang, T. Effect of Ta and Ni on γ' phase morphology and compression behavior of Co-8.8Al-9.8W superalloys. *Mater. Sci. Eng. A* **2022**, *839*, 142739–142748. [CrossRef]
35. Yao, X.; Ding, Q.; Wei, X.; Wang, J.; Zhang, Z.; Bei, H. The effects of key elements Re and Ru on the phase morphologies and microstructure in Ni-based single crystal superalloys. *J. Alloys Compd.* **2022**, *926*, 166835–166844. [CrossRef]
36. Yu, L.; Lu, Z.; Peng, S.; Li, X. Effect of Al/Ti ratio on γ' and oxide dispersion strengthening in Ni-based ODS superalloys. *Mater. Sci. Eng. A* **2022**, *845*, 143240–143253. [CrossRef]
37. Aghaie-Khafri, M.; Honarvar, F.; Zangeneh, S. Correlation between ultrasonic velocity and solutionising time in Rene 80 superalloy. *Mater. Sci. Technol.* **2011**, *27*, 1433–1435. [CrossRef]
38. Tang, Y.T.; Panwisawas, C.; Ghoussoub, J.N.; Gong, Y.; Clark, J.W.; Németh, A.A.; McCartney, D.G.; Reed, R.C. Alloys-by-design: Application to new superalloys for additive manufacturing. *Acta Mater.* **2021**, *202*, 417–436. [CrossRef]
39. Wu, B.; Liang, J.; Zhou, Y.; Yang, Y.; Li, J.; Sun, X. Effect of heat treatment on the microstructure and tensile properties of a new superalloy designed for additive manufacturing. *Mater. Sci. Eng. A* **2022**, *856*, 144023–144035. [CrossRef]
40. Taller, S.; Austin, T. Using post-processing heat treatments to elucidate precipitate strengthening of additively manufactured superalloy 718. *Addit. Manuf.* **2022**, *60*, 103280–103295. [CrossRef]
41. Chang, K.; Ma, L.; Li, P.; Lv, J.; You, X.; Zhang, Y.; Tan, Y. Effect of heat treatment on microstructure and mechanical properties of GH4099 superalloy fabricated by selective laser melting. *J. Alloys Compd.* **2023**, *934*, 167813–167826. [CrossRef]
42. Wang, Y.; Lei, L.; Shi, L.; Liang, F.; Wan, H.; Zhang, B.; Zhang, G. Effect of heat treatment on strain hardening ability of selective laser melted precipitation-hardened GH4169 superalloy. *Mater. Charact.* **2022**, *190*, 112064–112074. [CrossRef]
43. Zhang, L.; Li, Y.; Zhang, Q.; Zhang, S. Microstructure evolution, phase transformation and mechanical properties of IN738 superalloy fabricated by selective laser melting under different heat treatments. *Mater. Sci. Eng. A* **2022**, *844*, 142947. [CrossRef]
44. Lee, J.; Terner, M.; Jun, S.; Hong, H.-U.; Copin, E.; Lours, P. Heat treatments design for superior high-temperature tensile properties of Alloy 625 produced by selective laser melting. *Mater. Sci. Eng. A* **2020**, *790*, 139720. [CrossRef]
45. Chen, L.; Sun, Y.; Li, L.; Ren, X. Microstructural evolution and mechanical properties of selective laser melted a nickel-based superalloy after post treatment. *Mater. Sci. Eng. A* **2020**, *792*, 139649. [CrossRef]

46. Zhang, D.; Niu, W.; Cao, X.; Liu, Z. Effect of standard heat treatment on the microstructure and mechanical properties of selective laser melting manufactured Inconel 718 superalloy. *Mater. Sci. Eng. A* **2015**, *644*, 32–40. [CrossRef]
47. Wang, Y.; Li, N.; Jia, C.; Liu, W.; Wu, Y.; Gao, C.; Xiong, H. Effects of solution treatment on the microstructural evolution and tensile properties of GH 3230 superalloy manufactured by laser melting deposition. *Mater. Charact.* **2022**, *194*, 112339–112351. [CrossRef]
48. Wang, H.; Zhang, X.; Wang, G.; Shen, J.; Zhang, G.; Li, Y.; Yan, M. Selective laser melting of the hard-to-weld IN738LC superalloy: Efforts to mitigate defects and the resultant microstructural and mechanical properties. *J. Alloys Compd.* **2019**, *807*, 151662–151674. [CrossRef]
49. Smith, T.; Esser, B.; Antolin, N.; Viswanathan, G.; Hanlon, T.; Wessman, A.; Mourer, D.; Windl, W.; McComb, D.; Mills, M. Segregation and η phase formation along stacking faults during creep at intermediate temperatures in a Ni-based superalloy. *Acta Mater.* **2015**, *100*, 19–31. [CrossRef]
50. Kozar, R.; Suzuki, A.; Milligan, W.; Schirra, J.; Savage, M.; Pollock, T. Strengthening Mechanisms in Polycrystalline Multimodal Nickel-Base Superalloys. *Met. Mater. Trans. A* **2009**, *40*, 1588–1603. [CrossRef]
51. Zhang, H.; Li, Y.; Ma, T.; Chang, T.; Zhang, P.; Fang, X.; Huang, K. Tailoring of nanoscale γ' precipitates and unveiling their strengthening mechanisms in multimodal nickel-based superalloy GH4720Li. *Mater. Charact.* **2022**, *188*, 111918–111928. [CrossRef]
52. Gypen, L.A.; Deruyttere, A. Multi-component solid solution hardening. *J. Mater. Sci.* **1977**, *12*, 1034–1038. [CrossRef]
53. Roth, H.A.; Davis, C.L.; Thomson, R.C. Modeling solid solution strengthening in nickel alloys. *Met. Mater. Trans. A* **1997**, *28*, 1329–1335. [CrossRef]
54. Wen, H.; Tang, X.; Jin, J.; Cai, C.; Yang, H.; Teng, Q.; Wei, Q.; Wang, X.; Zheng, C.; He, Y.; et al. Effect of extrusion ratios on microstructure evolution and strengthening mechanisms of a novel P/M nickel-based superalloy. *Mater. Sci. Eng. A* **2022**, *847*, 143356–143372. [CrossRef]
55. Janez, G. The superalloys fundamentals and applications. *Int. J. Microstruct. Mater. Prop.* **2012**, *7*, 464–465.
56. Zhang, P.; Yuan, Y.; Shen, S.; Li, B.; Zhu, R.; Yang, G.; Song, X. Tensile deformation mechanisms at various temperatures in a new directionally solidified Ni-base superalloy. *J. Alloys Compd.* **2016**, *694*, 502–509. [CrossRef]
57. Fang, Q.; Li, L.; Li, J.; Wu, H.; Huang, Z.; Liu, B.; Liu, Y.; Liaw, P.K. A statistical theory of probability-dependent precipitation strengthening in metals and alloys. *J. Mech. Phys. Solids* **2018**, *122*, 177–189. [CrossRef]
58. Liu, F.; Wang, Z.; Tan, L.; Xiao, X.; Li, Y.; Li, X.; Zhang, H.; Zhu, L.; Jiang, L.; Liu, Y. Creep behaviors of fine-grained Ni-base powder metallurgy superalloys at elevated temperatures. *J. Alloys Compd.* **2021**, *867*, 158865. [CrossRef]
59. Crudden, D.; Mottura, A.; Warnken, N.; Raeisinia, B.; Reed, R. Modelling of the influence of alloy composition on flow stress in high-strength nickel-based superalloys. *Acta Mater.* **2014**, *75*, 356–370. [CrossRef]
60. Kruml, T.; Conforto, E.; Piccolo, B.L.; Caillard, D.; Martin, J.L. From dislocation cores to strength and work-hardening: A study of binary Ni3Al. *Acta Mater.* **2002**, *50*, 5091–5101. [CrossRef]
61. Galindo-Nava, E.I.; Connor, L.D.; Rae, C.M.F. On the prediction of the yield stress of unimodal and multimodal γ' Nickel-base superalloys. *Acta Mater.* **2015**, *98*, 377–390. [CrossRef]
62. Russell, K.; Brown, L. A dispersion strengthening model based on differing elastic moduli applied to the iron-copper system—ScienceDirect. *Acta Metall.* **1972**, *20*, 969–974. [CrossRef]
63. Jiao, Z.; Luan, J.; Zhang, Z.; Miller, M.; Ma, W.; Liu, C. Synergistic effects of Cu and Ni on nanoscale precipitation and mechanical properties of high-strength steels. *Acta Mater.* **2013**, *61*, 5996–6005. [CrossRef]
64. Kong, H.; Xu, C.; Bu, C.; Da, C.; Luan, J.; Jiao, Z.B.; Chen, G.; Liu, C. Hardening mechanisms and impact toughening of a high-strength steel containing low Ni and Cu additions. *Acta Mater.* **2019**, *172*, 150–160. [CrossRef]

materials

MDPI

Article

Structure and Mechanical Properties of Cu–Al–Mn Alloys Fabricated by Electron Beam Additive Manufacturing

Evgeny Moskvichev, Nikolay Shamarin and Alexey Smolin *

Institute of Strength Physics and Materials Science, Siberian Branch Russian Academy of Sciences, 2/4, pr. Akademicheskii, Tomsk 634055, Russia
* Correspondence: asmolin@ispms.ru

Abstract: In this work, the method of electron beam additive manufacturing (EBAM) was used to fabricate a Cu-based alloy possessing a shape memory effect. Electron beam additive technology is especially relevant for copper and its alloys since the process is carried out in a vacuum, which makes it possible to circumvent oxidation. The main purpose of the study was to establish the influence of the printing parameters on the structure of the obtained products, their phase composition, mechanical properties, dry friction behavior, and the structure-phase gradient that formed in Cu–Al–Mn alloy samples during electron beam layer-by-layer printing. The results of the study allowed us to reveal that the structure-phase composition, the mechanical properties, and the tribological performance of the fabricated material are mainly affected by the magnitude of heat input during electron beam additive printing of Cu–Al–Mn alloy. High heat input values led to the formation of the $\beta 1' + \alpha$ decomposed structure. Low heat input values enabled the suppression of decomposition and the formation of an ordered 1 structure. The microhardness values were distributed on a gradient from 2.0 to 2.75 GPa. Fabricated samples demonstrated different behaviors in friction and wear depending on their composition and structure, with the value of the friction coefficient lying in the range between 0.1 and 0.175.

Keywords: shape memory alloy; electron beam additive manufacturing; heat input; microstructure; mechanical properties; tribological performance

1. Introduction

Copper has good ductility, electrical conductivity, and thermal conductivity, which allows it to be used in a wide range of industries. In particular, copper-based alloys are traditionally used as materials for tribo-conjugations and friction units. Among all copper alloys, it is worth noting such an important class of materials as the shape memory alloys (SMA), the interest in which has recently increased [1–7].

In Cu–Al alloys, the β phase of the bcc A2 structure is formed directly from the melt. Furthermore, it undergoes an order–disorder transition: A2 (disordered bcc Cu) → B2 (CuAl) → $D0_3$ (Cu_3Al) [8]. Phase transformation-induced effects of shape memory and superelasticity consist of phase transitions from the high-temperature austenite phase to the low-temperature martensite phase and vice versa [9]. This is the main characteristic responsible for the properties of this class of alloys [10]. Phase transition temperatures for aluminum bronze are strictly dependent on the chemical composition of the alloy, as shown elsewhere [4,11–13].

Cu–Al–Mn alloys differ favorably from other copper-based SMA due to alloying with manganese. Manganese increases mechanical and corrosion properties and improves the technological characteristics of Cu–Al alloys. It increases not only strength, but also plasticity and, consequently, the ability for pressure-assisted machining. Bronze Cu–9Al–2Mn is well treated by pressure in both hot and cold conditions, whereas bronzes Cu–9Al–3Fe, Cu–10Al–3Fe–1Mn, and Cu–10Al–4Fe–4Ni are well deformed only in the hot state. At the same time, manganese significantly affects the formation of the beta phase, expanding the temperature range of its existence [9]. In tribology, martensitic transformations are considered new mechanisms of friction and structural adaptability of the material to loading conditions [14–16]. Dependence of the mechanical properties of these alloys on their structure was studied in [17,18]; to enhance these properties, it was suggested to use the addition of different elements.

Methods that are the most widely used for manufacturing Cu–Al–Mn bronze products include powder metallurgy and vacuum casting. At the same time, methods of local metallurgy and additive technologies are rarely represented in the literature in relation to such alloys, although they have wide possibilities for manufacturing materials with specific characteristics. Since these technologies allow varying the chemical composition of the material during layer-by-layer printing, they make it possible to create a structure-phase gradient with the required physical and mechanical properties in the finished product or semi-finished product directly in the local area exposed to load. This is especially topical for all Cu–Al alloys widely used in tribotechnical applications [19–21].

In this work, the method of electron beam additive manufacturing (EBAM) was used to fabricate wear resistant Cu–Al–Mn samples. Electron beam technology is particularly relevant for copper and its alloys since the process is carried out in a vacuum, which makes it possible to circumvent oxidation. The structure formation under the conditions of local fusing that characterize additive manufacturing is a complex process. However, additive manufacturing methods are now being increasingly used. As a result, understanding the structure formation in additive manufacturing technology has become a hot research topic. Due to the use of manganese aluminum bronze in tribo-conjugations, wear behavior analysis of fabricated samples also needs to be performed.

Because of their complex crystallization process, aluminum bronze alloys are very sensitive to technology and production parameters [9]. Recently, a large number of papers have appeared devoted to the additive manufacturing of nickel aluminum bronze [22,23], while much fewer studies have dealt with the additive manufacturing of manganese aluminum bronze.

The main purpose of the study was to establish the influence of the printing process parameters on the structure of the obtained products, their phase and chemical composition, mechanical properties, wear performance, and the structure-phase gradient that formed in Cu–Al–Mn alloy samples during electron beam layer-by-layer printing.

2. Materials and Methods

Commercial Cu–7Al aluminum bronze wire was selected for additive printing of the lower layers of the workpiece. Cu–Al bronze was chosen as the underlayer because it has a similar chemical composition, which excludes the formation of intermetallic compounds and the appearance of particles of a material other than aluminum bronze in the fusion zone of the layers of the underlayer material and Cu–Al–Mn layer. AISI 321 stainless steel was used as a substrate because it has a good ability to fuse with aluminum bronze. Substrate is a consumable material from which the finished product is cut off after printing is completed. Two Cu–Al–Mn alloys were used to print the upper layers; their chemical composition is given in Table 1. Cu–Al–Mn bronze was originally supplied in ingots manufactured by the standard metallurgical method. For additive printing, the ingots were cut into bars of the size of 3 × 3 × 20 mm, after which the surfaces of the bars were ground using sandpaper of 240 grit.

Table 1. Chemical composition of raw materials (wt.%).

Material	Cu	Al	Mn
Aluminum bronze (Cu–7Al)	93.5	6.3	0.2
Alloy 1 (Cu–11Al–9Mn)	79.2	11.2	9.6
Alloy 2 (Cu–11Al–4Mn)	85.1	11	3.9

For printing, a laboratory EBAM machine manufactured at the Institute of Strength Physics and Materials Science of the Siberian Branch of the Russian Academy of Sciences was utilized. This machine has been successfully used for 3D printing of various metal and alloy products with interesting properties that have been thoroughly studied and discussed elsewhere [24,25]. The main unit of the machine is an electron gun with a thermionic cathode in the form of a tantalum tablet with indirect heating. The accelerating voltage of this gun is 30 kV. The electron gun is equipped with a system for focusing and moving the electron beam. It is also possible to form various scans of the electron beam (ellipse, spiral, line, etc.). The printed material is usually fed in the form of a wire.

The printing process on the EBAM machine consists of the following steps:

- A substrate is installed on a copper water-cooled table.
- The focus of the electron beam is adjusted to the surface of the substrate.
- The wire feed mechanism is adjusted so that the wire falls into the area of the electron beam on the substrate (the wire is fed at an angle of ~45° to the table surface in the center of the beam).
- A working pressure (vacuum) is created.
- Printing is performed according to the specified algorithm for the formation of layers.

When printing, the electron beam forms a melt bath on the surface of the substrate/sublayer where the wire is fed, and the table is moved along three coordinates in accordance with the chosen printing strategy.

The preferred type of material transfer is continuous transfer, when the wire melts upon contact with the melt bath or its proximity without forming droplets. Drop transfer is less preferable because, in this case, it is hard to control the geometric shape of the product, and drops of molten filament can be greatly overheated, which in turn can lead to evaporation of some elements (in the presence of low-melting elements, a drop can spray without getting into the melt bath). Moreover, in some cases, radiation from the electron beam impact zone can repel droplets of molten filament.

Printing is carried out in a vacuum with a working pressure of 5 mPa. To remove excess heat energy during operation, the workpiece is formed on a substrate (in this work it is steel) located on a copper table with liquid cooling.

The bar method of material feeding during additive printing of Cu–Al–Mn layers was used in this work. This method has a higher productivity in comparison with wire feed and allows us to work with a wide range of materials, not limited to wire filaments. However, Cu–7Al layers were printed using the standard wire feed method. The printing process was arranged as follows: a Cu–7Al layer of the required height was applied to the steel substrate by using wire feed method. The wire feeder was removed from the table, a bar feeder was supplied, and further Cu–Al–Mn printing was carried out using bars. The image of the EBAM machine camera at the start of the process is shown in Figure 1.

The combined printing method described above makes it possible to create layers of material with a structure-phase gradient by controlling the temperature regime of printing and the depth of the mixing zone, and it also provides ample opportunities for creating layered structures with alternating layers of different materials.

The macrostructure of the obtained samples is depicted in the optical image of the typical cross-section shown in Figure 2a. The sample is fabricated in accordance with the specified printing strategy; therefore, the macrostructure of the cross-section is represented by crystallized melt pools 1 with a width of 3–4 mm (clearly seen in Cu–Al–Mn layers). This is because, in this work, a printing strategy with intermittent application of parallel layers was used (Figure 2b); when between passes, the sample was kept for a short time (10 s corresponding to dots in Figure 2b), required for changing the bar in the feeder, changing the printing track, and cooling the sample. As a result, boundaries were formed between adjacent tracks consisting of smaller grains crystallized at a higher speed. It is also worth noting the transition zone 3 in Figure 2a, formed at the border of the fusion of two bronzes.

Figure 1. Configuration of the camera of the EBAM machine for printing by a combined method with wire and bar feeding of the material: (1) wire feeder, (2) bar feeder, (3) cooled table, (4) steel substrate.

Figure 2. (**a**) Optical image of the bulk sample cross-section: (1) single track pool; (2) Cu–Al–Mn stable layers; (3) transition zone; (4) Cu–Al layers; (5) steel substrate. (**b**) Scheme of the printing strategy.

In the process of fabricating samples, the most optimal regime of printing and post-processing of the workpiece was selected to obtain the required parameters of the structure and chemical composition.

The calculation of heat input E during printing was carried out according to the following formula:

$$E = \frac{60 \times U \times I}{1000 \times V}, \tag{1}$$

where U is the voltage (kV), I is the current (mA), and V is the print speed (mm/min).

The selected parameters of the printing regimes are shown in Table 2.

Table 2. Selected parameters of the printing regimes.

Regime	Current I, mA	Voltage U, kV	Print Speed V, (mm/min)	Heat Input E, kJ/mm
0	40	30	50	1.44
1	40	30	100	0.72
2	40	30	200	0.36
3	40	30	300	0.24
4	40	30	400	0.18

When printing samples, it was found that the optimal regimes for obtaining intact Cu–Al–Mn samples were regimes 1–3 since printing in the regime 0 with high heat input led to an increase in the edge effect of the workpiece and, accordingly, its irregular shape. Printing in the low heat input regime, with high print speed (regime 4), led to the absence of fusion of bronze layers, as well as the formation of discontinuities and pores between the layers. After fabrication, some samples were subjected to homogenization annealing at a temperature of 900 °C for 4 h, followed by cooling into water.

Thus, hereinafter, we denote the fabricated samples as X-YAS, where X is the raw alloy (ingot) number in Table 1, and Y is the regime number in Table 2. The annealed samples are correspondingly denoted as X-YAN.

A field-emission scanning electron microscope (FE SEM) Tescan MIRA 3 LMU (TESCAN ORSAY HOLDING, Brno, Czech Republic) equipped with Oxford Instruments Ultim Max 40 EDS detector (Oxford Instruments, High Wycombe, UK) was used to analyze the structure of the materials fabricated in this work. The structure analysis was carried out with the following parameters: accelerating voltage, 20 kV; beam current, 1.6 nA. The analysis results were processed in AZtec v. 4.2 licensed software (Oxford Instruments, High Wycombe, UK). The samples for SEM analysis were cut in a direction perpendicular to the printing plane and were prepared by diamond polishing until 1 μm followed by silica polishing. The same samples were used later to measure microhardness.

The crystal structure of the samples was studied by X-ray diffraction using Shimadzu XRD-7000S (Shimadzu, Kyoto, Japan) in a Bragg–Brentano configuration using Cu–Kα1 radiation at 40 kV and 30 mA. The phase composition of the samples was identified using PDF-4+ 2015 software (ICDD, Newtown Township, PA, USA). The samples for XRD studies with dimensions of 10 × 10 mm were cut out in the printing plane from the uppermost layer; their surface was ground mechanically using sandpaper (400–2000 grit) followed by diamond polishing to 6 μm.

Microhardness was measured using a "Duramin-5" (Struers A/S, Ballerup, Denmark) microhardness tester at 100 g load. Wear tests were evaluated on a pin-on-disc tribometer (Tribotechnic, Clichy, France) operated at 0.1 m/s speed with a 20 N load for 90 min. Discs made of 440C steel were used as counterbodies.

3. Results and Discussion

3.1. Microstructure of the Obtained Samples

A set of EDX spectra were analyzed from square sections with a side of 250 μm and a step of 250 μm, starting from the substrate and finishing at the upper part of the sample. During the analysis, it was found out that the structure of the samples had a directional gradient of chemical composition (Figure 3). The manganese content depended on the value of the heat input and determined the structure-phase gradient formed in the samples. It is also worth noting the difference in the width of the gradient zone from the Cu–7Al substrate to the Cu–Al–Mn layers. In general, the gradient area could be divided into the following characteristic zones: (i) a zone of a sharp increase in the content of Mn; (ii) a zone of instability; (iii) a zone with a stable content of alloying elements. Annealing after printing significantly affected the structure of the gradient zone, leveling the chemical composition in the transition zone from Cu–7Al to Cu–Al–Mn. On the basis of the EDX analysis data, it was revealed that the pronounced transition zone formed during the printing and characterized by an uneven phase composition, as well as a reduced Mn content, was leveled during annealing, as can be seen in the plots in Figure 3. Due to the temperature treatment, it was possible to homogenize this area; however, in the case of Cu–11Al–9Mn alloy, this led to a blurring of the chemical composition of the transition zone (Figure 3).

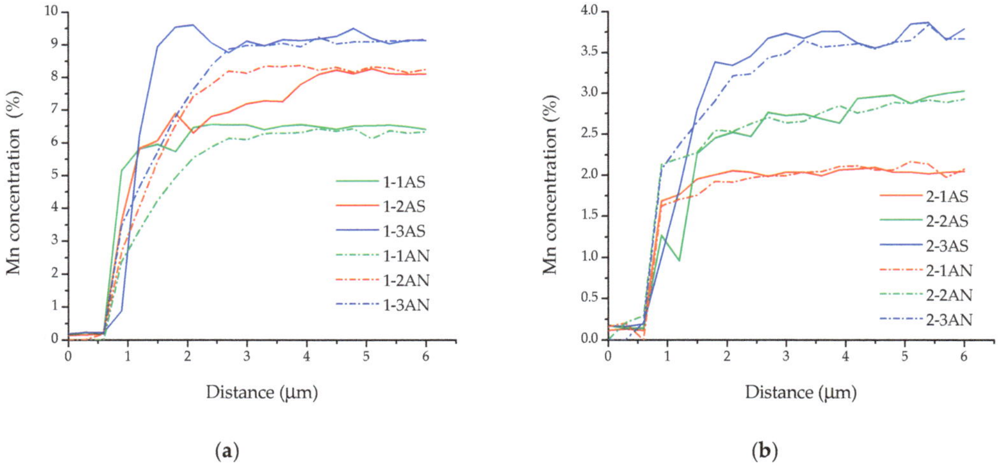

Figure 3. Concentration of manganese in the height of the samples in the state after printing (*X-Y*AS) and after annealing (*X-Y*AN): (**a**) fabricated from Alloy 1; (**b**) fabricated from Alloy 2.

It should be noted that, when printing in regimes with high heat input (1, 2), manganese burned out, whereas printing in regime 3 did not lead to significant losses (Figure 3). This effect is a well-known problem of temperature treatment or fabrication of metal materials.

Table 3 shows the data of quantitative EDX analysis collected from the top layer of samples. Sections of 250 × 250 μm were analyzed to certify the obtained material. The data in Table 3 show that a homogeneous layer of material was obtained at a distance of 2–3 mm from the Cu–7Al substrate as a result of the formation of a transition zone in the samples. This transition zone, in turn, was caused by mixing the deposited layers and substrate Cu–7Al layers.

Table 3. EDX analysis data (wt.%).

Sample	Cu	Al	Mn
1-1AS	83.0	10.6	6.4
1-2AS	80.9	10.9	8.2
1-3AS	79.6	11.1	9.3
2-1AS	87.1	10.8	2.1
2-2AS	86.1	10.8	3.1
2-3AS	85.5	10.8	3.7

Figure 4 depicts SEM micrographs of Cu–Al–Mn alloy samples immediately after printing in different regimes, and Figure 5 shows the same samples after annealing. The structure of Cu–11Al–4Mn alloy samples represents the consequences of the eutectoid decay of the β-phase. The ratio β/α depends on the printing regime since light elements burn out during the printing process, which also affects the width of the β-phase formation zone [26,27]. In the printing regime with low heat input (sample 2-3 in Figure 5f), crystallization occurs quickly due to the cooling of the table and the high print speed; the decay does not have time to occur sufficiently. The structure is mainly represented by the martensitic phase, with the formation of the α-phase only along the boundaries. Printing regimes with medium and high heat input lead to a greater degree of manganese burnout, greater heating of the sample and, as a result, a greater proportion of the formed grains of the α-phase (Figure 5d,e).

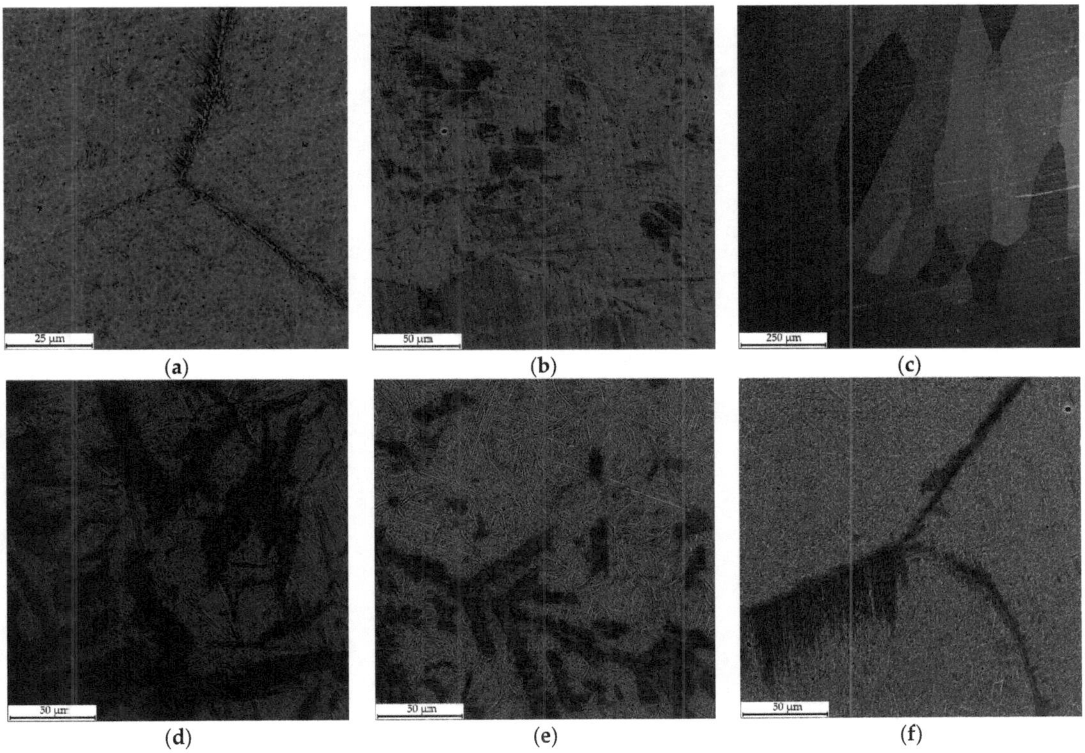

Figure 4. SEM-BSE images of as-printed samples microstructure: (**a**–**c**) for Alloy 1, 1–3 regimes, respectively; (**d**–**f**) for Alloy 2, 1–3 regimes, respectively.

Due to the initially higher amount of manganese in Cu–11Al–9Mn alloy samples, crystallization from the high-temperature β-phase occurred with the formation of an ordered β_1-phase [26,27]. Printing in a low-heat-input regime (samples 1–3 in Figure 5c) with quick cooling allowed suppressing eutectoid decomposition and obtaining a sample with large grains of the ordered β_1-phase (except the transition zone in which, due to mixing with Cu–7Al substrate, the alloy was depleted) [28,29]. Printing regimes with medium and high heat input led to the decomposition of the ordered β_1-phase and the formation of the $\alpha + \beta_1'$ system, similar to samples from Alloy 2 (Cu–11Al–4Mn). At the same time, the grain structure of sample 1-2 (Figure 5b) was characterized by both the presence of large grains characteristic of the ordered β_1-phase and a two-phase structure, as a consequence of the influence of the width of the transition zone.

The samples fabricated from Alloy 2 (Cu–11Al–4Mn) and subjected to annealing were characterized by a homogeneous needle-shaped martensitic structure with a well-defined interface between Cu–7Al layers and Cu–Al–Mn layers. After annealing, the samples fabricated from Alloy 1 (Cu–11Al–9Mn) had different structures depending on the printing regime. Thus, sample 1-1, printed with high heat input, transformed into martensite completely, sample 1-3, printed with low heat input, remained unchanged, and the intermediate sample 1-2, due to its wide transition zone, was divided into two parts, the lower of which (located closer to Cu–7Al) transformed into martensite, and the upper remained in the β_1 phase.

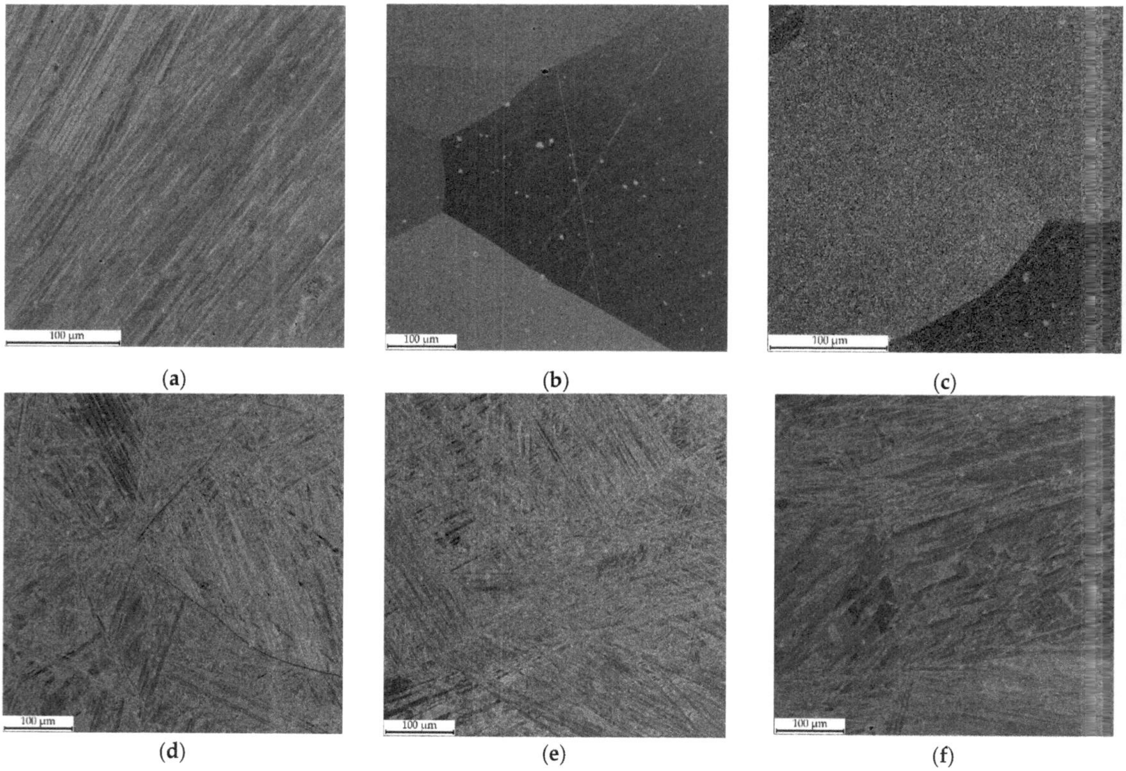

Figure 5. SEM-BSE images of annealed samples microstructure: (**a**–**c**) for Alloy 1, 1–3 regimes, respectively; (**d**–**f**) for Alloy 2, 1–3 regimes, respectively.

3.2. Microhardness of the Obtained Samples

Figure 6 presents the distributions of the microhardness along the height in all printed samples. According to an analysis of the data from Figures 3 and 6, the availability and width of the transition zone produced a structure-phase gradient that resulted in a gradient distribution of hardness in the samples. Samples printed from Alloy 1 (Cu–11Al–9Mn) were characterized by a specific zone of increased hardness (Figure 6a), a sharp increase in hardness in the transition zone, and its uniform distribution in the rest of the printed layer. Samples printed using bars from Alloy 2 (Cu–11Al–4Mn) were characterized by a smooth increase in hardness in the transition zone and an uneven distribution of hardness in the upper part of the samples (Figure 6b). This unevenness was due to the presence of a soft α-phase and a martensitic phase with high microhardness and excellent wear resistance [16]. A dependence of the width of the transition zone on the printing regime and heat input was also noted.

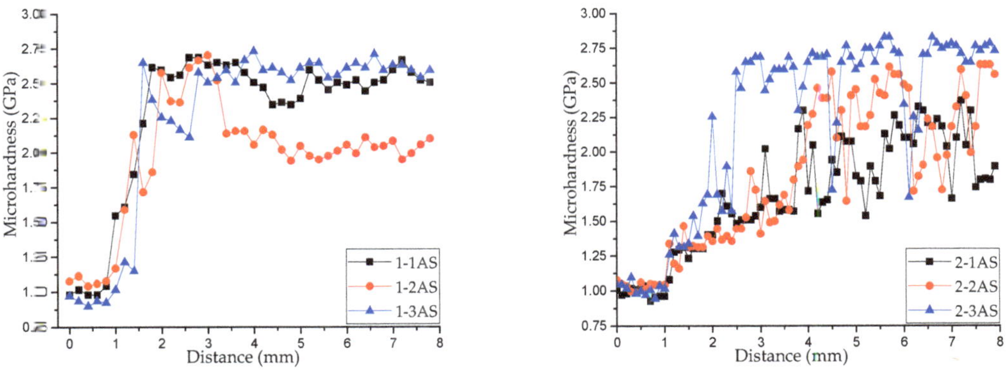

(a) (b)

Figure 6. Distribution of the microhardness over the height of the samples printed from Alloy 1 (**a**) and Alloy 2 (**b**).

Figure 7 presents the distributions of the microhardness along the height in the annealed samples. As can be seen from these plots, due to microstructural changes caused by annealing, the microhardness of the samples of series 2 and sample 1-1AN was within the same interval of values (2.0–2.4 GPa). This confirms the fact that annealing with quick cooling led to the transition of the structure to the martensitic phase, which can also be seen in Figure 5. Different properties of samples 1-2AN and 1-3AN can be explained by the fact that the transformation to the martensitic phase in them was restrained from the stable, ordered β_1-phase.

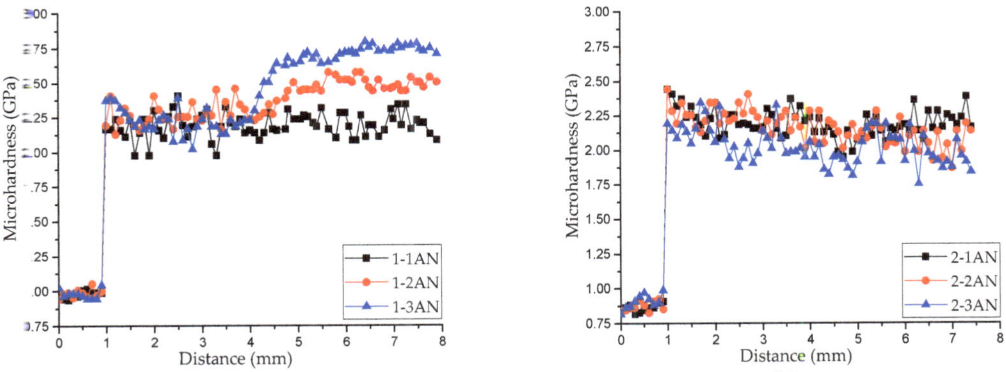

(a) (b)

Figure 7. Distribution of the microhardness over the height of the annealed samples printed from Alloy 1 (**a**) and Alloy 2 (**b**).

3.3. Wear Analysis

The results of pin-on-disc measurements are presented in Figure 8 and Table 4 for materials in different structural states. AS samples demonstrate a relatively low coefficient of friction in the range from 0.11 to 0.17. At the same time, the friction behavior of AS samples differs slightly among themselves. Because of the large amount of soft α-phase, sample 2-1AS had the lowest CoF, and sample 2-3AS had a CoF value between 2-1AS and samples 1-1AS and 1-3AS. However, all AS samples had a stable friction behavior with a short run-on stage and uniform CoF after process stabilization. The value of wear was also different for different printing conditions; however, in general, it was typical for aluminum bronze [30,31].

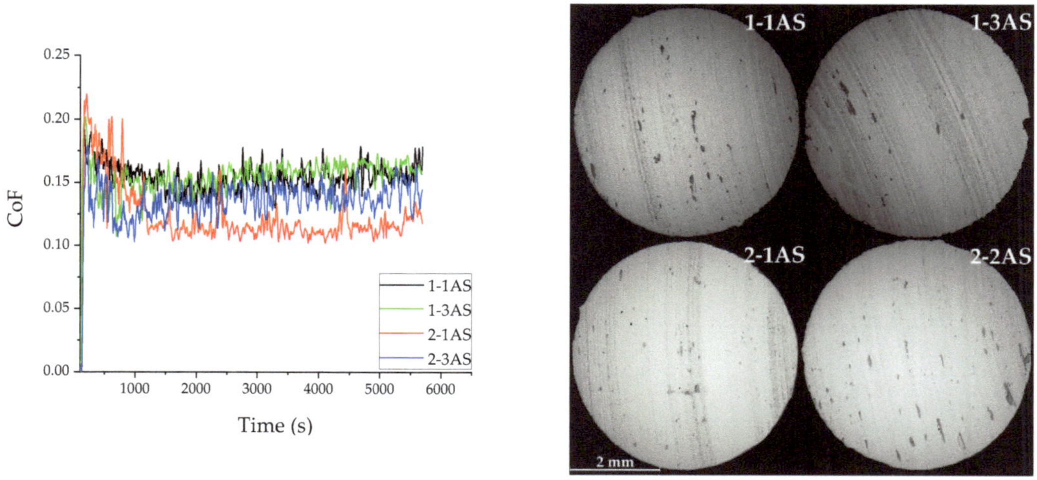

(**a**) (**b**)

Figure 8. Coefficient of friction versus time for different samples (**a**) and SEM-BSE images of wear surfaces (**b**).

Table 4. Wear rate of the fabricated samples in sliding friction.

Sample	Specific Wear Rate, $mm^3/(m \cdot N) \times 10^{-4}$
1-1AS	1.10
1-3AS	1.05
2-1AS	1.32
2-3AS	1.22

SEM analysis of friction surfaces (Figure 8b) showed a similar wear mechanism for all structure states. Oxides (the dark spots in Figure 8b) were formed during friction and were then crushed into smaller particles and smeared on the counterbody, forming a stable tribolayer. However, the difference in the structure and phase composition also led to uneven wear. For example, samples with a large amount of soft α-phase wore more quickly.

3.4. X-Ray Phase Analysis

Cu–11Al–9Mn alloy was characterized by two types of phase composition depending on the printing regime. The samples with high heat input underwent eutectoid decomposition, as evidenced by XRD data (Figure 9). Depending on the printing regime, the ordered β_1 phase either managed to complete the transition to $\alpha + \beta_1'$ (Figure 9c) or partially remained unchanged (Figure 9b). In the case of printing in the regime of the minimum permissible heat input (Figure 9a), the absence of Mn burnout and the high cooling rate made it possible to suppress decomposition, and the sample remained in the ordered β_1 phase [8,16,26,32].

As can be seen from Figure 10, the phase composition of the printed Cu–11Al–4Mn layers strongly depended on the printing regime. Thus, the Cu–11Al–4Mn alloy samples were characterized by a structure with the presence of the following two phases: R18-type martensite and α-phase. The appearance of the α-phase was caused by the eutectoid decomposition, which is typical for bronzes [33,34]. As can be seen from the comparison of SEM images (Figure 4) and XRD diagrams (Figure 10), in the low-heat-input regime, the cooling of the sample was faster and, consequently, less α-phase had time to separate. Accordingly, as evidenced by the high degree of intensity of X-ray peaks, mainly R18 martensite was present in the sample. Samples printed at high heat input values contained much more α-phase since high beam energies during printing contributed to high sample heating. From the XRD diagrams of the samples subjected to annealing (Figure 10b,d,f), it can be seen that the heat treatment allowed them completely transform into the martensitic phase.

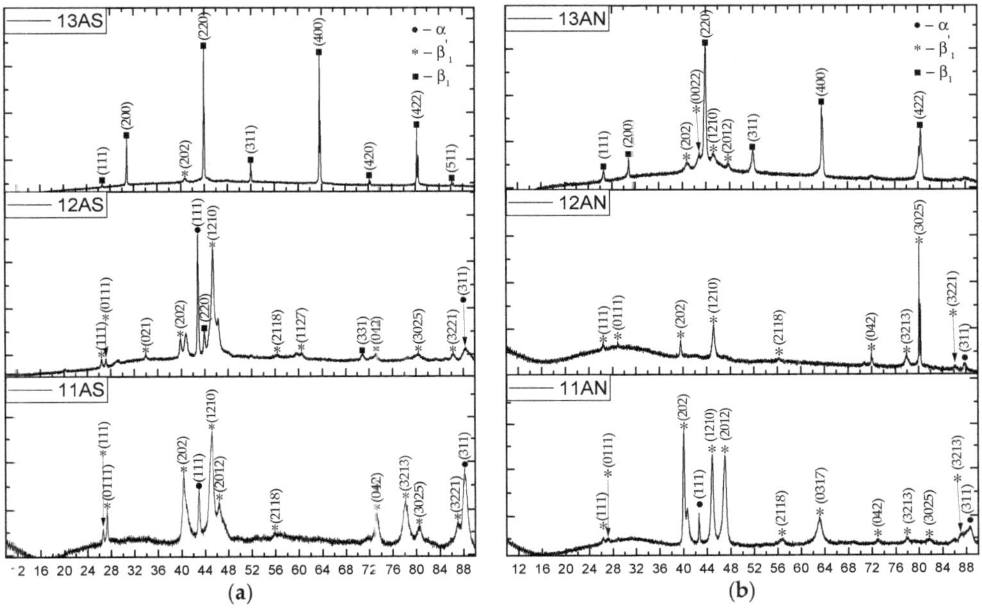

Figure 9. XRD diagrams of samples fabricated from Alloy 1: after printing (**a**); after annealing (**b**).

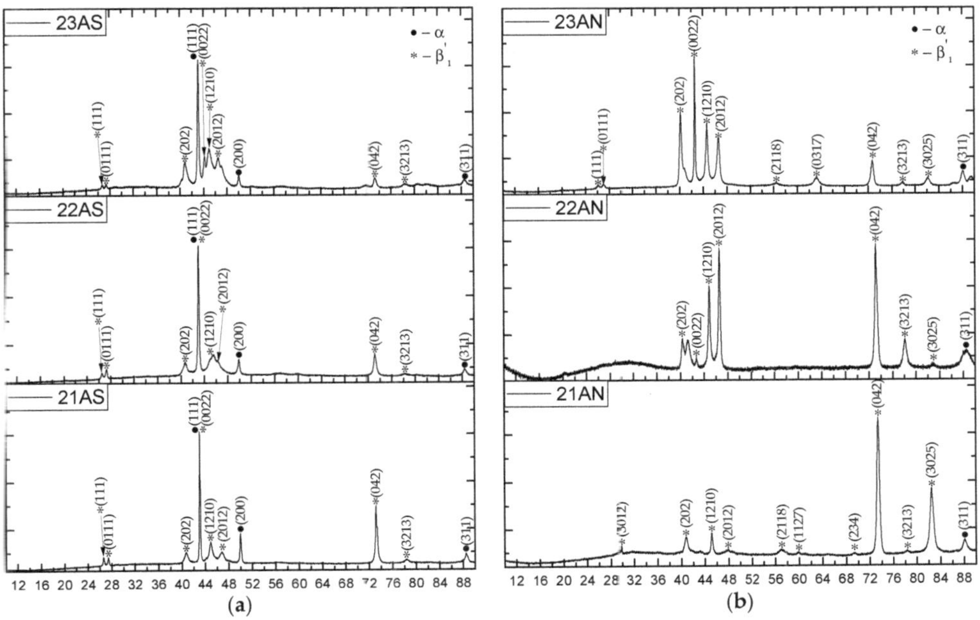

Figure 10. XRD diagrams of samples fabricated from Alloy 2: after printing (**a**), after annealing (**b**).

3.5. Discussion

The bulk samples from two alloys, Cu–11Al–4Mn and Cu–11Al–9Mn, were printed using the electron beam method at different heat input values. As an additional temperature treatment, some of the samples were annealed. The study showed a strict dependence of the structure-phase composition and mechanical characteristics on heat input during electron beam additive printing of the Cu–Al–Mn SMA alloys. It was established that the most optimal printing parameters corresponded to the value of heat input in the range from 0.24 kJ/mm to 0.72 kJ/mm, since they allowed obtaining a solid workpiece of the correct shape, without pores and melting defects, of controlled element-phase composition. Other authors also reported that alloys of aluminum bronze fabricated by different additive technologies with practically the same heat input had no pores or other defects [32,35,36]. The surface of the additively printed samples is usually not suitable for further practical use and must additionally postprocessed. For example, laser beam treatment can be used for the additional processing of the surfaces of the additively manufactured copper-based SMA [37].

The analysis of the microstructure by the SEM method showed a significant difference in the structure of samples printed with different heat input values. Thus, in the case of low heat input, the Cu–11Al–9Mn alloy sample remained in the ordered cubic $\beta_1{}'$-phase, and large grains were formed. It is worth noting that the same structure had the samples of Cu–Al–N—Mn bronze fabricated by selective laser melting in [32]. However, in the case of large values of heat input, the heat sink was insufficient, and the eutectoid decomposition had time to begin; the ordered $\beta_1{}'$-phase decomposed into α and β.

The gradient zone formed in the samples also depended on the heat input: the less heat input, the narrower the gradient zone. Subsequent heat treatment eliminated the difference, and the width of the gradient zone for all printing regimes was smoothed. The structure-phase gradient also caused an uneven distribution of mechanical characteristics in the areas of the printed workpiece. Thus, the presence of a soft α-phase led to an uneven distribution of microhardness even in the upper layers, whereas the availability of solid β_1 and $\beta_1{}'$-phases contributed to an increase in mechanical characteristics and consequently wear resistance. It is known that an increase in the amount of aluminum increases microhardness, as well as wear resistance [30]. However, the friction process and CoF mainly depend not only on hardness, but also on microstructure. That is why annealed martensitic samples possessing hardness lower than β_1 samples exhibited less wear. Intermetallic inclusions of solid phases also helped to significantly improve the characteristics of Cu–Al bronzes under dry friction [31].

Varying the heat input made it possible to control the structure, phase composition, and the structure-phase gradient. Thus, the variation of heat input when working with the raw material of Alloy 2 (Cu–11Al–4Mn) led to a significant change in the material. The formation zone of the β-phase narrowed due to changes in the chemical composition in the formed melt bath. In the process of further crystallization of the melt, the eutectoid decomposition of the β-phase into martensitic $\beta_1{}'$ and α occurred. Depending on the printing regime, the melt had time to decompose (in the case of high heat input) or did not have time (in the case of low heat input). The above analysis confirms that the effect of heat input was much higher and provided more possibilities for the studied material than for Cu–Si–Mn alloys [38].

When printing from the raw material of Alloy 1 (Cu–11Al–9Mn) in the low-heat-input regime, eutectoid decomposition did not occur at all during crystallization, allowing the fabrication of a workpiece with a typical bronze structure consisting of grains elongated in the direction of the temperature gradient. Printing in the regime of high heat input also led to excessive burnout of Mn and the beginning of eutectoid decomposition. Cu–Al–Mn alloys containing about 10 at.% Mn and 15–23 at.% Al were also characterized by ferromagnetic behavior, which was caused by the formation of the ferromagnetic β_3-phase of Cu_2AlMn composition [39].

The phase composition affected wear performance since samples with a soft phase showed a lower CoF value. In general, the nature of the wear behavior and the CoF value dynamics of printed samples were comparable with the literature data for bronze alloys [19,30]. By and large, the friction mechanisms of alloys possessing a shape memory effect are much more complex, including the formation of a structured tribolayer (wear particles and oxides from both the sample and the counterbody) and deformation processes in the near-surface layer [40]. Factors such as texture and adhesion transfer also play an important role, which were discussed in detail in [41,42]. A detailed assessment of all the factors influencing the friction of additively printed SMA Cu–Al–Mn bronzes can be the subject of a more extensive study and should undoubtedly be considered in future.

4. Conclusions

On the basis of the results obtained from this study, the following conclusions can be made:

1. We showed that the method of combined wire and bar feeding is promising for electron-beam additive manufacturing of Cu–Al–Mn shape memory alloys. Due to the variation of print regimes and filaments, samples with different structures and phase compositions were obtained. Print regimes with small heat input proved to be the most effective because they allowed both avoiding defects and producing the required chemical composition of the material.
2. Varying the heat input in the electron-beam additive manufacturing was evaluated on the basis of the structure analysis of the fabricated samples. It was established that crystallization occurred directly into the β-phase, which allowed obtaining the material without additional heat treatment in the case of raw material 1 (Cu–11Al–9Mn). In the case of raw material 2 (Cu–11Al–4Mn), heat treatment was necessary in any case. Thus, samples in a state characterized by phase transformation, which caused the shape memory effect, were obtained by 3D printing with a combined technology of material supply and subsequent heat treatment.
3. Structure analysis and analysis of the microhardness distribution allowed us to describe the structure-phase gradient formed in the fabricated samples during EBAM. Tribological tests allowed us to evaluate the effect of the print regime on the wear performance of the fabricated samples. Thus, the materials with a higher content of soft phase had better sliding friction characteristics, while the materials containing hard phase possessed worse ones. Creating a directional gradient in layer-by-layer printing (by varying the printing parameters) allows controlling the mechanical behavior of the resulting sample in the most stress-prone areas.

Author Contributions: Conceptualization, E.M., N.S. and A.S.; methodology, E.M. and N.S.; formal analysis, A.S.; investigation, E.M. and N.S.; resources, N.S.; data curation, N.S.; writing—original draft preparation, E.M. and N.S.; writing—review and editing, A.S.; supervision, A.S.; project administration, A.S.; funding acquisition, A.S. All authors have read and agreed to the published version of the manuscript.

Funding: The investigation was carried out with financial support from the Russian Science Foundation (Project No. 20-19-00743).

Institutional Review Board Statement: Not applicable.

Informed Consent Statement: Not applicable.

Data Availability Statement: The data that support the findings of this study are available from the corresponding author upon reasonable request.

Acknowledgments: The SEM analyses were carried out with the equipment of the Tomsk Regional Core Shared Research Facilities Center of National Research Tomsk State University. The center was supported by the Ministry of Science and Higher Education of the Russian Federation Grant No. 075-15-2021-693 (No. 13.RFC.21.0012)). The authors are grateful to Margarita Khimich for her help in XRD analysis.

Conflicts of Interest: The authors declare no conflict of interest.

References

1. Bae, J.; Nyamaa, O.; Lee, J.; Yun, S.; Woo, S.; Yang, J.; Kim, M.; Noh, J.; Si, T. Electrochemical properties of the Si thin-film anode deposited on Ti-Nb-Zr shape memory alloy in Li-ion batteries. *Electrochem. Commun.* **2022**, *140*, 107315. [CrossRef]
2. Lu, N.H.; Chen, C.H. Inhomogeneous martensitic transformation behavior and elastocaloric effect in a bicrystal Cu-Al-Mn shape memory alloy. *Mater. Sci. Eng. A* **2021**, *800*, 140386. [CrossRef]
3. Lalegani, M.; Bodaghi, M.; Serjouei, A.; Afazov, S.; Zolfagharian, A. Adaptive reversible composite-based shape memory alloy soft actuators. *Sens. Actuators A Phys.* **2022**, *345*, 113779. [CrossRef]
4. Santosh, S.; Kevin Thomas, J.; Rajkumar, K.; Sabareesh, A. Effect of Ni and Mn additions on the damping characteristics of Cu-Al-Fe based high temperature shape memory alloys. *J. Alloy. Compd.* **2022**, *924*, 166258.
5. Hong, H.; Gencturk, B.; Brown, S.A.; Hosseini, F.; Jain, A.; Aryan, H.; Saiidi, S.; Araki, Y.; Kise, S. Long-term corrosion resistance of Cu-Al-Mn superelastic alloys and steel rebar for use in bridges. *Constr. Build. Mater.* **2022**, *350*, 128795. [CrossRef]
6. Yang, H.; An, K.; Wang, Y.D. Anomalous high-temperature quasi-linear superelasticity of Ni-Fe-Ga-Co shape memory alloy. *J. Alloys. Compd.* **2022**, *909*, 164808. [CrossRef]
7. Trehern, W.; Ortiz-Ayala, R.; Atli, K.C.; Arroyave, R.; Karaman, I. Data-driven shape memory alloy discovery using artificial intelligence materials selection (AIMS) framework. *Acta Mater.* **2022**, *228*, 117751. [CrossRef]
8. Kainuma, R.; Satoh, N.; Liu, X.J.; Ohnuma, I.; Ishida, K. Phase equilibria and Heusler phase stability in the Cu-rich portion of the Cu–Al–Mn system. *J. Alloys. Compd.* **1998**, *266*, 191–200. [CrossRef]
9. Kainuma, R.; Takahashi, S.; Ishida, K. Ductile shape memory alloys of the Cu-Al-Mn system. *J. Phys. IV* **1995**, *5*, C8-961–C8-966. [CrossRef]
10. Aldırmaz, E.; Tataroğlu, A.; Dere, A.; Güler, M.; Güler, E.; Karabulut, A.; Yakuphanoglu, F. Cu-Al-Mn shape memory alloy based Schottky diode formed on Si. *Phys. B Condens. Matter* **2019**, *560*, 261–266. [CrossRef]
11. Trehern, W.; Ozcan, H.; Franco, B.; Hite, N.; Malone, N.; Loveall, B.; Morrison, T.D.; Benafan, O.; Karaman, I. Exploring thermomechanical functionality of CuAlMn as an extreme low temperature shape memory alloy. *Mater. Lett.* **2022**, *308*, 131216. [CrossRef]
12. López del Castillo, C.; Mellor, B.G.; Blázquez, M.L.; Gómez, C. The influence of composition and grain size on the martensitic transformation temperatures of CuAlMn shape memory alloys. *Scr. Mater.* **1987**, *21*, 1711–1716.
13. Raju, T.N.; Sampath, V. Effect of Ternary Addition of iron on shape memory characteristics of Cu-Al alloys. *J. Mater. Eng. Perform.* **2011**, *20*, 767–770. [CrossRef]
14. Levintant-Zayonts, N.; Starzynski, G.; Kopec, M.; Kucharski, S. Characterization of NiTi SMA in its unusual behaviour in wear tests. *Tribol. Int.* **2019**, *137*, 313–323. [CrossRef]
15. Sattari, M.; Kadkhodaei, M.; Akbarzadeh, S.; Gholami, R.; Beheshti, A. Wear in superelastic shape memory alloys: A thermomechanical analysis. *Wear* **2022**, *488–489*, 204139. [CrossRef]
16. Tao, X.P.; Zhang, S.; Zhang, C.H.; Wu, C.L.; Chen, J.; Abdullah, A.O. Effect of Fe and Ni contents on microstructure and wear resistance of aluminum bronze coatings on 316 stainless steel by laser cladding. *Surf. Coat. Technol.* **2018**, *342*, 76–84. [CrossRef]
17. Jiao, Z.; Yin, F.; Wang, Q.; Hao, G.; Zhang, J.; Liu, L.; Ji, P.; Cui, C. Refining effect of an intermetallic inoculant on a Cu–Al–Mn shape memory alloy. *Mater. Chem. Phys.* **2022**, *280*, 125835. [CrossRef]
18. Pandey, A.; Hussain, S.; Nair, P.; Dasgupta, R. Influence of niobium and silver on mechanical properties and shape memory behavior of Cu-12Al-4Mn alloys. *J. Alloys. Compd.* **2020**, *836*, 155266. [CrossRef]
19. Shaik, M.A.; Golla, B.R. Densification, microstructure and properties of mechanically alloyed and hot-pressed Cu–15 Wt% Al alloy. *J. Mater. Sci.* **2018**, *53*, 14694–14712. [CrossRef]
20. Akram, S.; Babutskyi, A.; Chrysanthou, A.; Montalvão, D.; Whiting, M.J.; Pizurova, N. Improvement of the wear resistance of nickel-aluminium bronze and 2014-T6 aluminium alloy by application of alternating magnetic field treatment. *Wear* **2021**, *480–481*, 203940. [CrossRef]
21. Liu, X.; Huang, D.; Yan, C.; Zhou, Y.; Yan, W. Multi-directional forging and aging treatment effects on friction and wear characterization of aluminium-bronze alloy. *Mater. Charact.* **2020**, *167*, 110511. [CrossRef]
22. Dharmendra, C.; Hadadzadeh, A.; Amirkhiz, B.S.; Janaki Ram, G.D.; Mohammadi, M. Microstructural evolution and mechanical behavior of nickel aluminum bronze Cu-9Al-4Fe-4Ni-1Mn fabricated through wire-arc additive manufacturing. *Addit. Manuf.* **2019**, *30*, 100872. [CrossRef]
23. Shakil, S.I.; Dharmendra, C.; Amirkhiz, B.S.; Verma, D.; Mohammadi, M.; Haghshenas, M. Micromechanical characterization of wire-arc additive manufactured and cast nickel aluminum bronze: Ambient and intermediate temperatures. *Mater. Sci. Eng. A* **2020**, *792*, 139773. [CrossRef]
24. Tarasov, S.Y.; Filippov, A.V.; Shamarin, N.N.; Fortuna, S.V.; Maier, G.G.; Kolubaev, E.A. Microstructural evolution and chemical corrosion of electron beam wire-feed additively manufactured AISI 304 stainless steel. *J. Alloy. Compd.* **2019**, *803*, 364–370. [CrossRef]
25. Zykova, A.; Chumaevskii, A.; Vorontsov, A.; Kalashnikov, K.; Gurianov, D.; Gusarova, A.; Kolubaev, E. Evolution of microstructure and properties of Fe-Cu, manufactured by electron beam additive manufacturing with subsequent friction stir processing. *Mater. Lett.* **2022**, *307*, 131023. [CrossRef]

26. Kainuma, R.; Takahashi, S.; Ishida, K. Thermoelastic martensite and shape memory effect in ductile Cu-Al-Mn alloys. *Metall. Mater Trans. A* **1996**, *27*, 2187–2195. [CrossRef]
27. Sutou, Y.; Koeda, N.; Omori, T.; Kainuma, R.; Ishida, K. Effects of ageing on bainitic and thermally induced martensitic transformations in ductile Cu–Al–Mn-based shape memory alloys. *Acta Mater.* **2009**, *57*, 5748–5758. [CrossRef]
28. Oliveira, J.P.; Panton, B.; Zeng, Z.; Omori, T.; Zhou, Y.; Miranda, R.M.; Braz Fernandes, F.M. Laser welded superelastic Cu–Al–Mn shape memory alloy wires. *Mater. Des.* **2016**, *90*, 122–128. [CrossRef]
29. Mallik, U.S.; Sampath, V. Influence of aluminum and manganese concentration on the shape memory characteristics of Cu–Al–Mn shape memory alloys. *J. Alloys. Compd.* **2008**, *459*, 142–147. [CrossRef]
30. Shaik, M.A.; Golla, B.R. Development of highly wear resistant Cu-Al alloys processed via powder metallurgy. *Tribol. Int.* **2019**, *136*, 127–139. [CrossRef]
31. Kucita, P.; Wang, S.C.; Li, W.S.; Cook, R.B.; Starink, M.J. The effects of substrate dilution on the microstructure and wear resistance of PTA Cu-Al-Fe aluminium bronze coatings. *Wear* **2019**, *440–441*, 203102. [CrossRef]
32. Gustmann, T.; dos Santos, J.M.; Gargarella, P.; Kühn, U.; Van Humbeeck, J.; Pauly, S. Properties of Cu-based shape-memory alloys prepared by selective laser melting. *Shape Mem. Superelasticity* **2016**, *3*, 24–36. [CrossRef]
33. Oliveira, J.P.; Crispim, B.; Zeng, Z.; Omori, T.; Braz Fernandes, F.M.; Miranda, R.M. Microstructure and mechanical properties of gas tungsten arc welded Cu-Al-Mn shape memory alloy rods. *J. Mater. Proc. Technol.* **2019**, *271*, 93–100. [CrossRef]
34. Grgurić, T.H.; Manasijević, D.; Kožuh, S.; Ivanić, I.; Anžel, I.; Kosec, B.; Bizjak, M.; Bajsić, E.G.; Balanović, L.; Gojić, M. The effect of the processing parameters on the martensitic transformation of Cu-Al-Mn shape memory alloy. *J. Alloy. Compd.* **2018**, *765*, 664–676. [CrossRef]
35. Shen, C.; Pan, Z.; Ding, D.; Yuan, L.; Nie, N.; Wang, Y.; Luo, D.; Cuiuri, D.; van Duin, S.; Li, H. The influence of post-production heat treatment on the multi-directional properties of nickel-aluminum bronze alloy fabricated using wire-arc additive manufacturing process. *Addit. Manuf.* **2018**, *23*, 411–421. [CrossRef]
36. Dharmendra, C.; Shakerin, S.; Ram, G.D.J.; Mohammadi, M. Wire-arc additive manufacturing of nickel aluminum bronze/stainless steel hybrid parts—Interfacial characterization, prospects, and problems. *Materialia* **2020**, *13*, 100834. [CrossRef]
37. Santosh, S.; Kevin Thomas, J.; Pavithran, M.; Nithyanandh, G.; Ashwath, J. An experimental analysis on the influence of CO_2 laser machining parameters on a copper-based shape memory alloy. *Opt. Laser Technol.* **2022**, *153*, 108210. [CrossRef]
38. Filippov, A.; Shamarin, N.; Moskvichev, E.; Savchenko, N.; Kolubaev, E.; Khoroshko, E.; Tarasov, S. The effect of heat input, annealing, and deformation treatment on structure and mechanical properties of electron beam additive manufactured (EBAM) silicon bronze. *Materials* **2022**, *15*, 3209. [CrossRef]
39. Silva, L.S.; Silva, R.A.G. Alloys-by-design: Role of atomic properties on the phase equilibria of CuAlMn-based alloys. *Mater. Charact.* **2020**, *163*, 110304. [CrossRef]
40. Blau, P.J. Transitions in the sliding wear mechanisms of binary Cu-Al alloys having a range of solute contents and microstructures. *Wear* **2021**, *476*, 203681. [CrossRef]
41. Tarasov, S.Y.; Filippov, A.V.; Kolubaev, E.A.; Kalashnikova, T.A. Adhesion transfer in sliding a steel ball against an aluminum alloy. *Tribol. Int.* **2017**, *115*, 191–198. [CrossRef]
42. Tarasov, S.Y.; Chumaevskii, A.V.; Lychagin, D.V.; Nikonov, A.Y.; Dmitriev, A.I. Subsurface structural evolution and wear lip formation on copper single crystals under unlubricated sliding conditions. *Wear* **2018**, *410–411*, 210–221. [CrossRef]

materials

Article

Programmable Density of Laser Additive Manufactured Parts by Considering an Inverse Problem

Mika León Altmann [1,*], Stefan Bosse [2], Christian Werner [1], Rainer Fechte-Heinen [1,3,4] and Anastasiya Toenjes [1,3]

1 Leibniz-Institute for Materials Engineering-IWT, Badgasteiner Str. 3, 28359 Bremen, Germany
2 Department of Mathematics and Computer Science, University of Bremen, Bibliothekstr. 5, 28359 Bremen, Germany
3 Faculty of Production Engineering, University of Bremen, Bibliothekstr. 1, 28359 Bremen, Germany
4 MAPEX Centre for Materials and Processes, University of Bremen, 28359 Bremen, Germany
* Correspondence: altmann@iwt-bremen.de; Tel.: +49-421-218-51414

Abstract: In this Article, the targeted adjustment of the relative density of laser additive manufactured components made of AlSi10Mg is considered. The interest in demand-oriented process parameters is steadily increasing. Thus, shorter process times and lower unit costs can be achieved with decreasing component densities. Especially when hot isostatic pressing is considered as a post-processing step. In order to be able to generate process parameters automatically, a model hypothesis is learned via artificial neural networks (ANN) for a density range from 70% to almost 100%, based on a synthetic dataset with equally distributed process parameters and a statistical test series with 256 full factorial combined instances. This allows the achievable relative density to be predicted from given process parameters. Based on the best model, a database approach and supervised training of concatenated ANNs are developed to solve the inverse parameter prediction problem for a target density. In this way, it is possible to generate a parameter prediction model for the high-dimensional result space through constraints that are shown with synthetic test data sets. The presented concatenated ANN model is able to reproduce the origin distribution. The relative density of synthetic data can be predicted with an R^2-value of 0.98. The mean build rate can be increased by 12% with the formulation of a hint during the backward model training. The application of the experimental data shows increased fuzziness related to the big data gaps and a small number of instances. For practical use, this algorithm could be trained on increased data sets and can be expanded by properties such as surface quality, residual stress, or mechanical strength. With knowledge of the necessary (mechanical) properties of the components, the model can be used to generate appropriate process parameters. This way, the processing time and the amount of scrap parts can be reduced.

Keywords: additive manufacturing; LPBF; PBF-LB/M; AlSi10Mg; machine learning; inverse problem; adjustable relative density; demand-oriented process parameters

1. Introduction

Due to the possibility of the nearly unconstrained freedom of design and the low amount of wasted material the additive manufacturing process spreads into many industries. Further, the production costs as well as the production time can be decreased using additive manufacturing techniques [1]. One commonly known process is the laser powder bed fusion process (LPBF) also known as powder bed fusion of metal with a laser beam (PBF-LB) or selective laser melting (SLM), a powder-based laser additive manufacturing process [2,3]. Further industrial-relevant processes are electron beam melting (EBM) and laser metal deposition (LMD) [4]. Parts built in these types of processes consist of multiple layers erected one after another [2]. In three steps, the whole process can be described: Pre-process (preparation of the 3D data), manufacturing process (building the part), and post-process (heat treatment and surface finishing) [2]. The cost of additive manufactured parts is mainly driven by the machining time and the material costs. Thus, the needed machine time accounts to more than half of the total costs [5]. This depends on the build rate (BR), further on the time per layer, and the layer thickness [5,6]. State-of-the-art process parameters are chosen to reach the highest possible relative density and mechanical properties [2]. The relative density highly depends on the process parameters as well as the powders quality. During the process, pores arise due to spontaneous interruptions of the melt tracks and not entirely molten powder particles or gassed material. Commonly, the induced energy density (ED) is used to appraise the resulting density without considering physical effects such as the depth of the melt track and the prior mechanisms [7]. The ED is an indication if the material exhibits unmolten areas or even gassed ones. These pores, especially with sharp contours resulting from low EDs, can act as a starting point for cracks and reduce the reachable mechanical properties [8]. As shown by Bertoli et al. [7] for 316 L, the same ED values can be obtained with different process parameter combinations. Further, each process parameter influences the resulting relative density reducing the significance of the ED and can obtain misleading conclusions. Thus, the BR is subordinated to the resulting relative density, which causes too high safety factors for less stressed parts and part areas. Especially, if no high mechanical requirements exist [9]. Parts with densities higher than 99% have comparable mechanical properties to conventional casted or milled parts which are not always needed [2]. In the process, relative densities over 99% can become obsolete if hot isostatic pressing (HIP) is applied afterwards. Hereby the pores resulting from the LPBF process are closed by applying temperature and pressure. At a critical temperature, the external pressure exceeds the yield strength of the printed material and consequently leads to densification. Thus, parts with relative densities of 95% can be densified to 99.8% opening a wide range for improving the BR of the LPBF process [10]. This way, Herzog et al. [10] were able to increase the scan speed by 2/3 with slightly higher residual stresses. Through demand-oriented process parameters, the BR can be increased for less important part areas allowing significantly lower machine times. Demand-oriented process parameters for parts or part areas are current research topics. This way, e.g., more complex geometries can build with homogenous properties (density, surface roughness) due to locally different heat transfer [11]. For practical use the LPBF manufactured parts need specific properties which can be set by the process parameters adequately [12]. One approach for process parameter prediction is presented by Park et al. [13].

High relative densities can be reached by a variety of significantly different process parameters [7]. Bai et al. show that the hatch distance is a key factor when the layer thickness is constant. This way, a variety of laser powers and scan speeds can be chosen to reach the same density values. Small hatch distances further improve the mechanical properties of the parts [14]. Increasing the layer thickness from 40 µm to 50 µm the manufacturing process is much faster with nearly identical mechanical properties [15]. Increasing the layer thickness has no effect on the final phases of 18Ni-300 (MS1 maraging Steel), while changing other process parameters results in austenite volume changes [15]. To reach targeted relative densities, the parameters for the LPBF process have to be chosen. Large experimental designs are needed to discover the resulting densities because each parameter (continuously adjustable within the limits of the LPBF machine specification) as well as the interactions influences the results [2]. Predicting the relative density for known parameters is constrained by the complexity of the interactions of the parameters. Machine learning models can be trained to solve this problem allowing more precise predictions of the resulting relative density [13,16]. However, they also need a great amount of experimental data. This could be reasonable in fact that the models can be adapted for different materials or further mechanical properties with decreasing amounts of data needed. Additionally, machine learning algorithms are able to learn physical connections only from data, enabling them to learn very complex regressions and connections without the need for a priori assumptions [17]. Minbashian et al. [17] showed that artificial neural network (ANN) models obtain accuracies such as the most complex multiple regression models with a high amount of included terms. The inverse problem must be taken into account when generating parameters automatically [13,18]. As a result of the complex interactions, a nearly infinite amount of possible parameter combinations for one specific density value exists [14]. Common machine learning algorithms are not able to learn a problem with a higher dimension of outputs than inputs [18]. One solution could be recommendation algorithms used for choosing from an outsized number of possibilities. Most of them return recommendations by a single domain, e.g., the relative density. Yu et al. presented a support vector machine approach for a multi-domain recommendation algorithm [19]. By adding bound vectors, support vector machines can be improved [20]. By adapted and concatenated machine learning methods, this mathematically underdetermined inverse problem can be solved by taking up the forward model [13,18,21]. This concatenation is based on the structure of generative adversarial networks (GAN), in which a discriminator classifies data sets generated by a generator as real or synthetic. Both models are trained together, so that after the training is completed, artificial data is generated which can no longer be distinguished from real data. In this context, GANs belong to unsupervised learning methods [22]. By concatenating the backward model with the trained forward model, the model can be trained and supervised to a given target value. By modifying the loss function, hints can be used to improve the generated parameters with respect to the BR.

From a materials engineering perspective, the LPBF process can produce a high number of specimens in a relatively short time. The amount of data points is restricted by the subsequent metallographic processes. These processes include grinding, polishing, and microscopic imaging to provide metallurgical micrographs of each specimen. However, from a machine learning perspective, the amount of provided data set is relatively small. Therefore, the used ANNs are relatively compact, which should be able to cope with a smaller amount of data for the training process. This paper is a first serve of applying machine learning models in material sciences and deriving a deeper process understanding within the field of LPBF. Therefore, a statistical test series represented by a full factorial design should create the fundamental data set for this paper. This way, it should be examined how well ANNs perform on small statistical test series in predicting material properties.

In the following, the work is split into four sections. First, the inverse problems with data-driven methods are described. The Materials and Methods section includes specimen manufacturing and analysis, process parameter selection and evaluation, data analyses, and the model-building process used in this work. Afterwards, the results for a theoretic and real data-driven model are presented and discussed. Finally, conclusions are drawn. It is shown that LPBF components made of AlSi10Mg can be manufactured with a defined relative density. For this purpose, the inverse problem is solved.

2. Inverse Problems with Data-Driven Methods

Inverse problems can be found in a wide range of science and engineering applications, with the aim to infer an unknown quantity F that is not accessible (observable) directly. There is only access to another observable quantity G that is related to it via a linear or a non-linear relation H [23], such that $G = H(F)$. Solving inverse problems with hidden, complex, or partially unknown physics is often expensive and requires different formulations and simulations [24]. Inverse problems are mainly based on regularization theory and commonly use discretized representations of the unknown model function [23]. Discretization introduces an error ε, such that $G = H(F) + \varepsilon$. Probabilistic Bayesian-inference-based methods are used for solving inverse problems, too, especially if the observable measurement data is noisy, uncertain, or some are missing and with some outliers. A principle discussion of regularization, Bayesian methods, and the benefit of machine learning can be found in [23]. In addition to Bayesian optimization, finding an optimum of a function F without any assumptions about the form of F, in [25], generative methods and global optimization are proposed to solve inverse problems (such as material design), based on data-driven methods.

Simulation-based approaches, additionally combined with experimental data, using, e.g., physical informed neural networks (PINN), are suitable to solve inverse problems [24,26,27], but the required physical differential equations are currently not accessible for such a complex material process considered in this work, or at least they would oversimplify the problem in a way making this approach unsuitable for process parameter predictions.

Typical data-driven modeling maps a set of observables x as input variables on feature variables y as the output. The model function $f(x)$: $x \to y$ derived from measured (or simulated) data contains hidden system parameters of interest, e.g., process parameters, which can be contained in x for generalization. The system parameters of interest cannot be measured directly [21], and moreover, the output features are inductive knowledge retrieved by inference, but in methodological science, deductive knowledge is required. For process optimization, e.g., the system parameters given for a specific feature y are of interest, defining an inverse problem $f^{-1}(y)$: $y \to x$. The system and input parameters x are typically metric variables, whereas the output feature variables can be metric or categorical variables.

A model function $f(x)$: $x \to y$ is bijective if it is invertible and there is a (unique) bidirectional mapping of $x \Leftrightarrow y$, i.e., for each x there is a y, and for each y there is an x, e.g., $f(x) = x + 1$. Linear functions are commonly invertible. In the case of multivariate functions, a set of equations is necessary to solve the inversion analytically. A solver that inverts a function f should assess the diversity of possible inverse solutions (if not unique) for a given measurement and should be able to estimate the complete posterior distribution of the system parameter [21].

Complete inversion with unique solutions is a challenge, and even a simple perceptron (artificial neuron) cannot be inverted without additional constraints or auxiliary variables:

$$\begin{aligned}
y &= f\left(\vec{x}\right) = g\left(\sum_i x_i \cdot w_i + b\right) \\
y &= s\left(\vec{x}\right) = \sum_i x_i \cdot w_i \to s^{-1} = \left\{ x_i = \frac{y - b - \sum_{j \neq i} x_j w_j}{w_i} \right\} \\
y &= g(x) = k \cdot x \to g^{-1} = \tfrac{y}{k} \\
y &= g(x) = \tfrac{1}{1+e^{-x}} \to g^{-1} = \ln\left(\tfrac{y}{1-y}\right)
\end{aligned} \tag{1}$$

Here w_i are internal model parameters (not to be confused with system parameters) and g is the transfer function, e.g., a linear, rectified linear, or the saturating sigmoid-logistics function. Inversion of functions can introduce poles and high non-linearities. The transfer function can be commonly fully or partially invertible, but the inversion of the sum results in a set of equations with an infinite set of solutions. The main reason for this non-solvable inversion problem is the information and dimensionality reduction in a perceptron ($\mathbb{R}^n \to \mathbb{R}$). The strong information reduction is typical for ANN creating an ill-posed problem space. This under-determined problem can be solved by using multiple combined and overlapped perceptrons, but requires still that the dimensions of x and y are equal, i.e., $|x| = |y|$.

With respect to data-driven and experimental methods, the forward path $x \to y$ is easily accessible, but the backward path $y \to x$ is initially hidden. Using invertible functional graph networks, these networks can be trained by the data-driven approach. Invertible neural networks (INN) satisfy three major features: The mapping between input and output is bijective, both forward and backward computations are efficient, and direct computation of the posterior probabilities is possible [13]. To address the information loss in many data-driven problems, a latent and auxiliary output variable z is introduced normalizing the input and output dimensions, i.e., $|x| = |y| + |z|$ and $x \leftrightarrow [y,z]$ become a bijective mapping. The well-posed forward path can be trained and supervised, instead of the ill-posed inverse problem, see Figure 1b. It is required that the latent variable z is independent of y, and should follow an easy distribution, e.g., $N(0,1)$. The inverse model is basically a conditional and tractable probability $p(x \mid y)$, but the derivation from the forward model is not tractable (being just an approximation by the latent probability distribution $p(z)$ and $z \in \mathbb{R}^k$ and multi-variate normal distribution N). The latent variable distribution is derived in a training process by, e.g., the maximum mean discrepancy (MMD), which is a kernel-based method for comparison of two probability distributions that are only accessible through samples [21]. To train such an $x/y/z$ network, the supervised data loss L_y and the unsupervised latent space loss L_z must be considered in the optimization process. If both losses reach zero, $g = f^{-1}$ returns the true posterior $p(x,y)$ for any y. For improved convergence, an additional loss L_x on the input side can be used and implemented again by MMD [21]. So, INNs are basically auto-encoders whose codes have the same size as the original data.

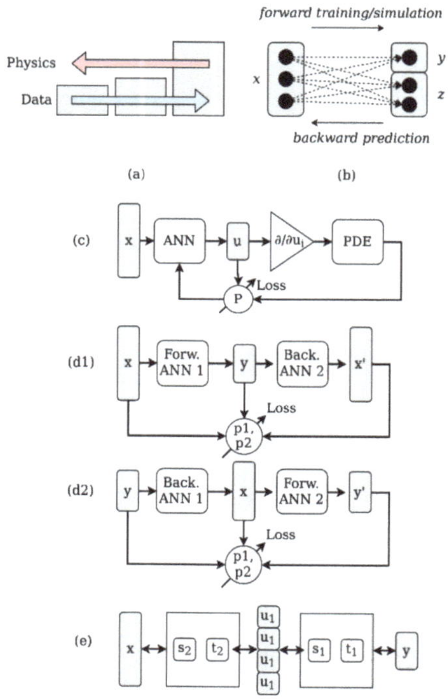

Figure 1. (**a**) Decreasing correlation of physics with increasing volume of data; (**b**) solving the ambiguous inversion problem by adding latent variable z; (**c**) basic concept of physics-informed ANN training; (**d1,2**) forward and backward model chain with closed-loop training; (**e**) affine coupling layer embedding four ANN s and t.

An invertible network is typically composed of two independent but coupled ANN, one for the forward and one for the backward path. A simple chain of the forward and backward model is similar to the auto-encoder, whereas in [28] an affine coupling layer is used to embed multiple ANN models for the forward and backward paths by using latent variable pairs $[u_1, v_1]$ and $[u_2, v_2]$. Every single ANN must not be invertible. Both architectures are compared in Figure 1d,e. The training process, such as in the proposed method in this work, is a closed-loop iterative process incorporating the forward and backward model with its specific loss functions as described above.

However, even if we have a suitable invertible architecture, e.g., an invertible neural network, training of deep neural networks requires a big amount of data in terms of diversity, variance, noise, and completeness of the input and output space, not always available for scientific problems such as the one addressed in this work [24]. Additionally, an increase in the data volume can result in a decrease in physical correlation, see Figure 1a. So, even solving the forward path can be challenging, and forward models derived from incomplete and sparse data sets lack the required generalization and are especially important for the inversion. Additional constraints derived from physical laws can support the training and optimization process by limiting the hyper-parameter space of the optimization problem (minimizing the overall $|y_o - y|$ error). Physics-informed learning integrates (noisy) data and mathematical models and implements them through neural networks or other kernel-based regression networks [24]. Both, the information loss of ANN and the contradiction of data amount and correlation with physics characterize the problem to be solved in this work, shown in Figure 1.

A lot of physical models are related to partial differential equations (PDE). The basic concept, as an extension of the INN, is the combination of an ANN outputting intermediate (code) features that are passed to a PDE, which is part of the training loss path, see Figure 1c. Although the data set used in this work poses strong sparseness, parameter gaps, bias, and noise, physics-informed methods are not applied here due to the (unknown) complex and probabilistic material models arising in additive manufacturing processes. Even simulation of the additive manufacturing process is a challenge with respect to computational complexity and matching real-world physics.

3. Materials and Methods

3.1. LPBF and Metallography

The base of the examination is a statistical experiment design with a full factorial combination of four steps per parameter printed on an SLM 125 HL (SLM Solutions Group, Lübeck, Germany) in common aluminum alloy AlSi10Mg, see Table 1. The maximum laser power used for this work is set at 350 W. Parameter programming and print file preparation were performed in Materialise Magics (Materialise GmbH, Bremen, Germany). The influence of the laser power P_L, the hatch distance h_S, the scan speed v_S, and the layer thickness D_S are investigated. The layer thicknesses are chosen so that two-layer thicknesses can be printed in one job by using the skip layer function, which reduces the amount of build jobs.

Table 1. Factors and steps for the statistical experiment design without heated build plate.

Factor	Step I	Step II	Step III	Step IV
D_S in μm	40	50	80	100
P_L in W	100	180	260	340
v_S in mm/s	500	1150	1800	2450
h_S in mm	0.05	0.08	0.11	0.14

The rest of the process parameters were chosen in even steps constrained by minimum and maximum values representing extreme parameter selections. In this way, it should be possible to achieve both very low and very high relative densities. The EDs studied range from 2.92 J/mm^3 to 340 J/mm^3. All relative densities are measured by image analysis of metallurgical micrographs. To speed up the analytical process, three process parameter sets are combined in one specimen. It can be guessed that the three segments in each specimen are mutually influenced through heat conduction. However, it can be observed in Figure 2, that there is no significant influence of the heat conduction when using multiple parameters in one geometry. The selected region of interest (ROI) in the core of each segment for relative density measurement should be a representative value without the influence of thermal conduction or contour parameters. Additionally, the process parameters in each specimen are randomized to avoid spatial dependencies due to heat conduction.

This way, 128 process parameter combinations can be printed in 43 specimens. By using the skip layer function (layer thicknesses of 40 μm and 80 μm, as well as 50 μm and 100 μm, are printed in one run), only two print jobs with 43 specimens each are needed to produce 256 combinations. After printing, the specimens are embedded in a polymer, ground, and polished parallel to the build direction before micrographs are taken. Thus, lower relative densities can only be measured with an increasing error caused by the removal of unsolidified powder particles during grinding.

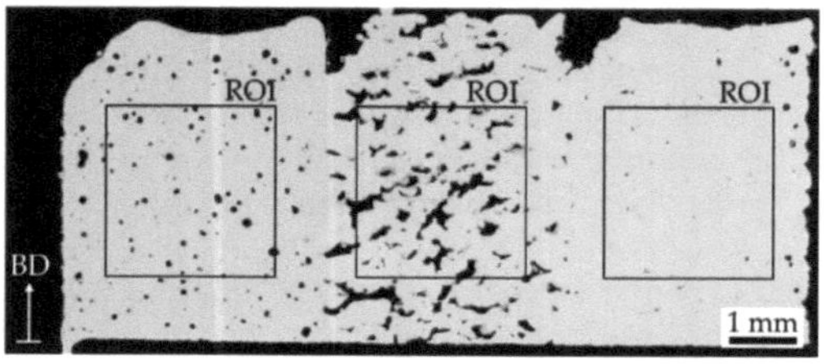

Figure 2. Metallurgical micrograph of three different parameter sets and densities combined in one specimen geometry with a layer thickness of 80 µm and marked region of interest (ROI), as well as the build direction (BD) and the used process parameters.

3.2. Data Analysis and Preparation

The most common use for machine learning algorithms is the identification of patterns in large data sets or regression functions where, e.g., no physical laws are known [29]. Conversely this means that a sufficiently large number of data points is needed for the training process. For this reason, it is necessary to extend the aforementioned data set with results from other projects, resulting in more than 400 instances. Differences in the AlSi10Mg powder used as well as the particle distribution are not considered in this work. Afterwards, statistical methods are applied to identify the data sets' quality which is linked to the achievable model quality [30]. Based on these results, the data set will be prepared for the model training process. Removing missing data and invalid as well as valid outliers should improve the quality of the trained models [16,31]. Missing data points can result from process parameters that do not allow the fabrication of dense specimens that can be analyzed by micrographs. Further invalid outliers are, e.g., process parameter values that cannot be set in the used machine or very low relative densities that cannot be reliably measured with the methods used. Valid outliers will be identified by a quartile analysis [32]. Knowing valid outliers, the range of interesting relative densities can be set. It can be assumed that relative densities below 70% are not technologically useful.

As common, the net data set will be scaled linearly into the same value interval for all attributes and targets afterwards [33]. The small net number of instances (350) is increased by normal distributed fuzziness for the process parameters. A simple function for the standard deviation with respect to the relative density is established by a quick check of the process stability. This allows the data set to be expanded to 700 instances.

3.3. Modeling

For the modeling process, ANNs are prepared using the python framework PyTorch [34]. Due to the dynamic graphs in PyTorch, loss values can be manipulated before they are backpropagated through the model. To check if the forward and especially the backward modeling problem can be solved by concatenated ANNs and manipulating the loss value, artificial data will be used. Afterwards the models can be trained on the real data described before.

As mentioned, the forward modeling problem and the inverse modeling problem, see Figure 3, are addressed in this article. To solve the forward prediction problem, resulting in relative density for known process parameters, common supervised ANNs [35] will be trained on the prepared database. As usual, the input dimension is higher than the output dimension. Predicting the relative density resulting from a given set of LPBF process parameters is fundamental for the inverse modeling process. Thus, the influence of each process parameter needs to be modeled to be able to solve the inverse problem. This is condensed in the following hypothesis.

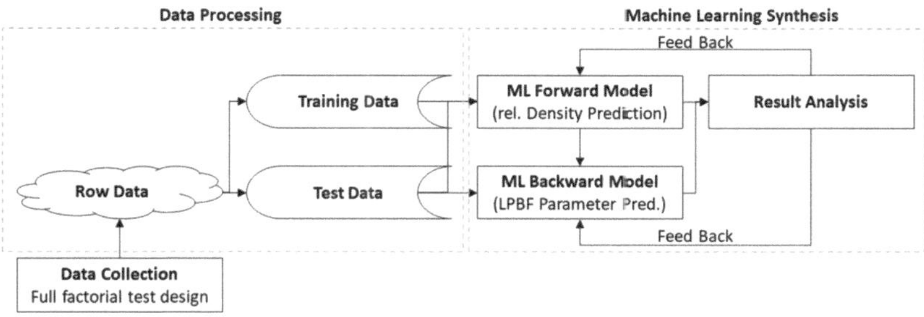

Figure 3. Data flow diagram for the training of the forward (relative density prediction) and the backward model (LPBF process parameter prediction) used in this work.

Hypothesis 1. *By the use of machine learning algorithms, the resulting relative density of given process parameters can be predicted even no direct correlation exists.*

Solving the inverse problem, process parameter prediction for a target density should firstly be achieved by creating a database with the best trained forward model as shown in Figure 4 and introduced by Park et al. [13]. Each factor is subdivided into steps regarding the value interval defined in Table 1. With a full factorial combination, a parameter space with up to 100 k combinations will be created. Through the trained forward model, the densities of each combination can be predicted. Detecting the best parameter sets occurs by searching algorithms and the build rate as a constraint.

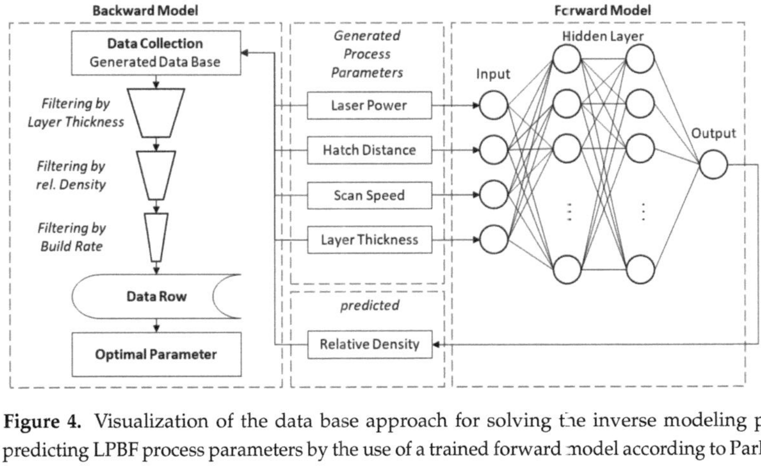

Figure 4. Visualization of the data base approach for solving the inverse modeling problem of predicting LPBF process parameters by the use of a trained forward model according to Park et al. [13].

Hypothesis 2.1. *With the trained forward model, a database can be created that contains a finite amount of possible solutions for the backward modeling problem. Using a search algorithm and constrained by the build rate, process parameters for specific density values, and layer thicknesses can be identified.*

Secondly, the inverse modeling problem can be solved by concatenating the forward model with an inverse ANN architecture, see Figure 5. This principle is oriented to the functionality of GANs. The inverse ANN generates a set of process parameters. Subsequently, the trained forward model predicts the resulting density. This manipulation allows a comparison of the target density and the resulting one. In this step, the forward model acts as a simple calculation in which operations are written to the dynamic graph of the variable. By following the dynamic graph backwards, the gradients are set and the loss value can be unfolded. This allows the regular not trainable inverse problem to be trained.

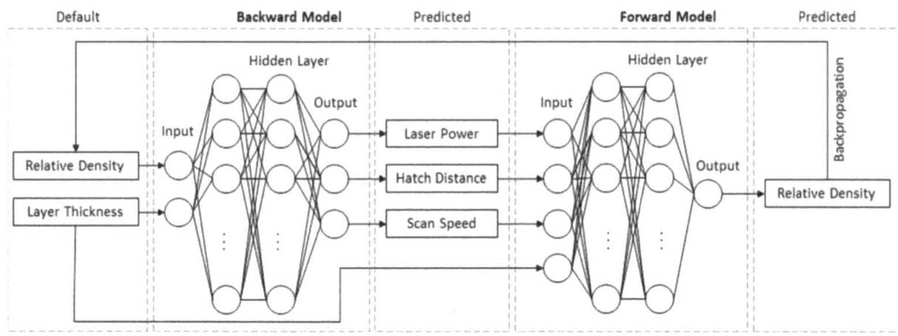

Figure 5. Visualization of the concatenated ANN models for the forward modeling problem (rel. density prediction) and the inverse modeling problem (LPBF process parameter prediction) with in- and outputs used for training and inference.

Hypothesis 2.2. *By the use of the dynamic graphs in PyTorch, an inverse architecture can be concatenated with a trained forward model. This way, it is possible to train an ANN with a lower dimension of the inputs than the outputs. Furthermore, the process parameters can be optimized regarding the achieved build rate by adding a hint to the loss function.*

4. Results and Discussion

4.1. Examination of the LPBF Process

The experimental test series shows the process limits at low as well as high EDs. Overall, 242 out of 256 parameter combinations are possible to build. Those that cannot be built have ED values < 15 J/mm^3 and track energies < 1 J/mm. Overall densities in the range of 20% to nearly 100% were produced. Three classes of pore formations can be spotted. In Figure 6 are metallurgic micrographs mapped by the laser power and scan speed for two-layer thicknesses and three hatch distances. First, prior unshaped pores can be detected accompanied by low relative densities, colored blue in Figure 6. Second, there are big prior round pores accompanied by medium to high relative densities, colored grey in Figure 6. Third, there are micrographs with small round and occasional unshaped pores reaching the highest mean density, colored orange in Figure 6. By decreasing the volume energy (lower laser power and higher scan speeds), big unshaped pores of the blue class occur. At very high-volume energies (low scan speed with high laser power), big round pores of the grey class result. Decreasing the hatch distance increases the number of orange classified as high-density results at the expense of the blue class, see Figure 6. By increasing the layer thickness from 40 µm to 100 µm, more process parameters with lower laser power cannot be built up. Moreover, the amount of orange classified micrographs decreases significantly, see Figure 6.

The three classes of micrographs depend on the shape of the pores correlating to the pore-building mechanisms. The prior unshaped pores (blue) are caused by the lack of a fusion mechanism. The introduced energy is too low for melting the layers entirely, resulting in poor connections between the layers [36]. Large and prior round pores (grey) result from the keyhole effect. Thus, the energy introduced is high enough to gas/gasify the material and a long melt pool lifetime allows small pores to merge into large round pores [37]. Between both the conduction mode (orange) as the sweet spot for the highest possible relative densities is located. According to Wang et al., the short lifetime of the melt tracks allows only large gas pores to evacuate [37]. Small and round pores are characteristic of this mechanism [37]. If the energy is locally too low, occasional unshaped pores results.

The process window for reaching high relative densities gets smaller with an increase in the hatch distance as well as the layer thickness. Further, occasional unshaped pores become more likely at higher layer thickness. This can indicate an increased sensitivity for process parameter fluctuations at higher layer thicknesses.

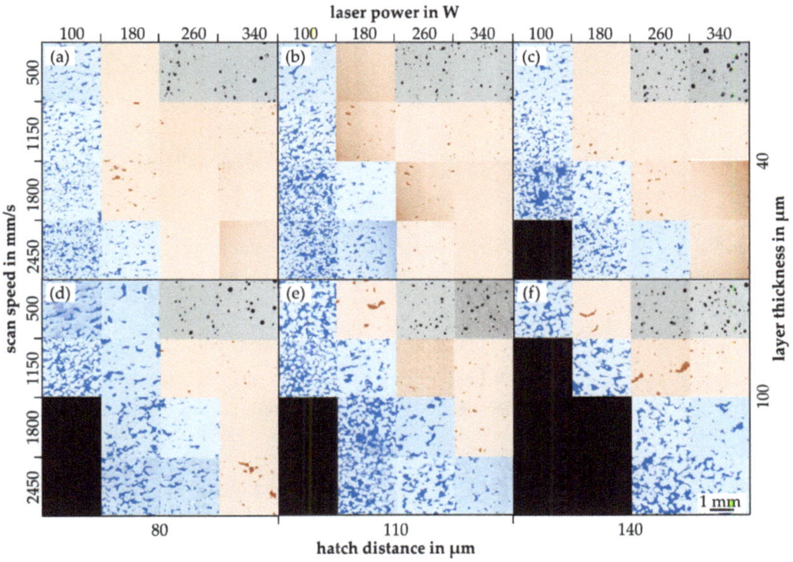

Figure 6. Mapping of metallurgical micrographs for different process parameters colored depending on the pore building mechanism (blue: lack of fusion, orange: conduction, grey: keyhole): micrographs by the laser power and scan speed for a hatch distance and laser power of (**a**) 80 µm and 40 µm; (**b**) 110 µm and 40 µm; (**c**) 140 µm and 40 µm; (**d**) 80 µm and 100 µm; (**e**) 110 µm and 100 µm; (**f**) 140 µm and 100 µm.

Correlations between the process parameters with each other and with the relative density are not detectable. A wide spectrum of relative densities can be reached with all four-factor steps. Further, all factor steps of the full factorial experiment design are clearly visible in the correlation plots as columns of data points parallel to the y-axis, see Figure 7a,b. Between the relative density and the ED, a correlation such as an exponential saturation can be detected, see Figure 7c. It should be noted that the majority of the process parameter combinations result in relative densities above 90%. This can be explained by the high thermal conductivity of the aluminum alloy (AlSi10Mg) used and the low melting temperature range. Thus, a high relative density can already be achieved with relatively low introduced ED values. In order to introduce higher porosities, the energy must be greatly reduced, which further increases process instability. This process instability results not only in a higher porosity but also in a higher standard deviation. The fact that a large number of relative densities can be achieved with a single ED may indicate a significant influence of the selected process parameters and their interactions, see Figure 7d. At this point, the ED can be confirmed as a key parameter for a rough estimation of the achievable relative density for a process parameter set [6]. However, further interactions have to be considered and investigated in detail.

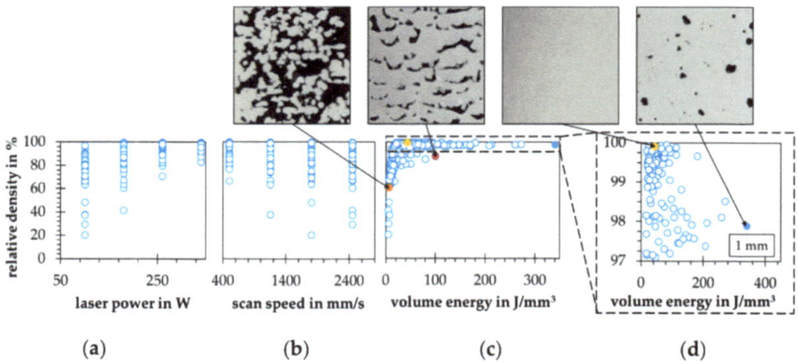

Figure 7. Correlation plots for the relative density depending on selected process parameters resulting from the process parameter examination (**a**) by the laser power; (**b**) by the scan speed; (**c**) by the volume energy (ED); (**d**) by the volume energy (ED) in detail.

4.2. Data Analysis

The database contains 388 instances after adding data from previous experiments (null values excluded), the four process parameters as well as the ED and the resulting relative densities. The distribution of each variable becomes apparent in the histograms, see Figure 8.

Mainly, the right-side hanging distribution of the layer thickness, the laser power, and the ED get into focus, see Figure 8a–e. The relative density exhibits a left-side hanging distribution, see Figure 8e. These formations indicate that the variables with low densities are underrepresented in the database. This can affect the machine learning training negatively, because missing points cannot flow into the regression model. Complementary, the density plot of the relative density is a sign of a wide range of possible process parameter combinations reaching relative densities above 95%. This can become critical for solving the hypothesis of predicting process parameters for target densities primarily in the case of reaching lower density values. Further, the full factorial experiment design conditionals only four parameter steps for each variable. Even with additional data, the cardinality of the four process parameters is in the area of categorical variables, restricting the reachable model quality further. So, the cardinalities are between six (layer thickness) and 44 (scan speed). The volume energy as well as the relative density own cardinalities of 343 and 279. Constrained by the possible parameter space of the used SLM 125 HL invalid outliers with laser powers > 350 W are removed. Further, density values < 70% and hatch distances > 0.25 mm are detected as valid outliers and are removed too. Thus, a database with 350 instances results.

4.3. Forward Modeling Problem-Density Prediction

First, a simple regression model for calculating the resulting relative density by the ED according to Figure 7c is created, see Figure 9a. Using this model, a set of artificial equally distributed process parameters and relative densities is built. For the model training, 280 instances and for the test 120 instances are used. By this, an ANN with 4 inputs and 1 output neuron with two hidden layers including 8 and 4 neurons is trained. The neurons of all layers are activated by a sigmoid function. By use of the Adam algorithm [38] and the mean squared error loss [39], a test prediction error of 0.17% (0.12% for training data) and an R^2-value of 0.98 (for training and test data) can be obtained, see Figure 9b. If using the ReLU activation function for the hidden layers, a much higher error occurs, see Figure 9c.

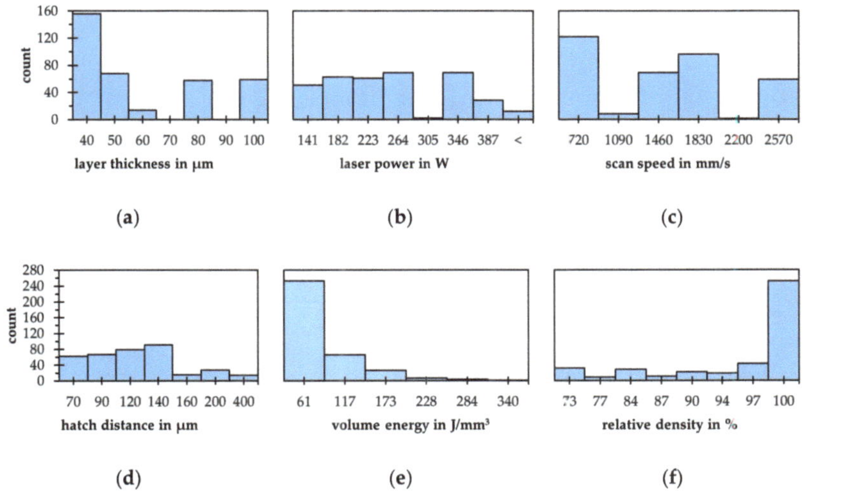

Figure 8. Histograms of all process parameters and the relative densities of the gross data base with 388 Instances (**a**) layer thickness; (**b**) laser power; (**c**) scan speed; (**d**) hatch distance; (**e**) volume energy density; (**f**) relative density.

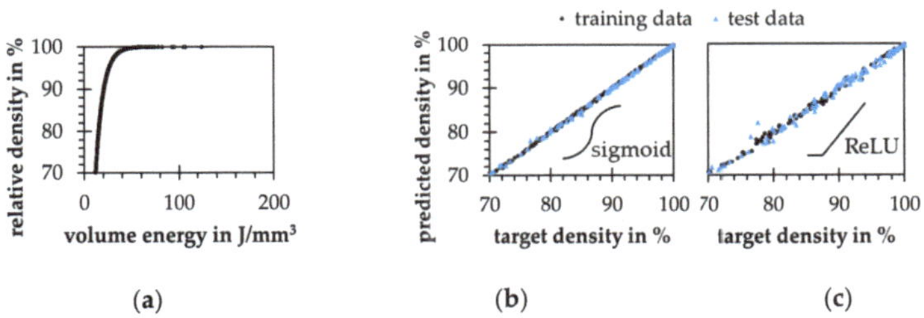

Figure 9. (**a**) artificial relative density over the volume energy density (ED) resulting from the regression model; predicted density over target density of the trained forward model with use of the artificial data (**b**) with sigmoid; (**c**) and ReLU activation.

Proof of Hypothesis 1. At this point, hypothesis 1 can be confirmed regarding the artificial data. By the use of machine learning algorithms, especially ANNs, the relative density can be predicted by LPBF process parameters. □

4.4. Backward Modeling Problem-Process Parameter Prediction

The first approach mentioned in hypothesis 2.1 is a database. The trained forward model is used to predict relative density values for an incremental varied process parameter space. Each parameter is increased in 9 and 18 steps from the minimal to the maximal value. Once the database is calculated, process parameters can be selected by the relative density, the layer thickness, and the BR. The predicted process parameters reach the targeted relative densities, see Figure 10c, because all combinations of the database are synthesized by the use of the trained forward model. The prediction error of the forward model influences the quality of both backward models. The database is also able to rebuild the ED distribution regarding the test data, see Figure 10d. In comparison to the ANN model, the EDs are slightly higher, compare Figure 10d,b. The mean BRs are 3% to 30% lower than the ones of the ANN model. Increasing the number of parameter steps also increases the mean BR since more possible solutions can be considered.

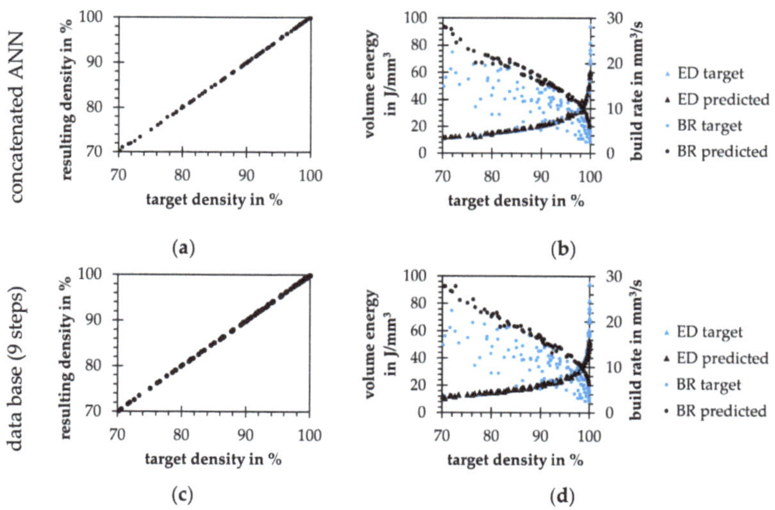

Figure 10. (**a**) resulting densities from predicted process parameters predicted by the trained forward model; (**b**) original and predicted volume energy (ED) and the build rate (BR) distribution over the target density; (**c**) resulting densities over target densities; (**d**) original and predicted ED and BR distributions over the target density.

Secondly, for the prediction of process parameters according to hypothesis 2.2, a relative density, as well as a desired layer thickness, are used as inputs. The laser power, scan speed, and hatch distance will be predicted. For this, a model architecture of 2 input neurons, 2 hidden layers with 4 and 8 neurons, and 3 output neurons is chosen. With the use of artificial data, the concatenated model can learn the inverse problem in less than 10 k training cycles. As shown in Figure 10a, the process parameters predicted by the inverse model reach the targeted densities with a prediction error of 0.09%. Additionally, the model is able to depict the relatively simple ED distribution of the artificial test data set, see Figure 10b. The prediction error depends also on the quality of the chosen forward model. In contrast to approaches such as INNs [21] or non-dominated sorting genetic algorithm (NSGA-II) and modified variational autoenconder (MVAE) [18], the concatenated ANN model does not unfold the complete possible result space. For every pair of target density and target layer thickness, only one process parameter prediction exists. The mentioned models, also the shown database approach [13], unfold the whole result space, so constraints are needed to reduce the number of possible solutions. These constraints can be included in the concatenated ANN training reducing the computing time for process parameter prediction.

For a constant layer thickness of 0.05 mm, a mean BR for relative densities between 70% and 99% of 18.91 mm^3/s is reached. By adding a hint to the error function of the model, training the build rate can be increased by 12.5% up to 20.45 mm^3/s. Two types of hints are formulated, a linear (*MLin*) and a squared (*MSQ*):

$$MSQ = \frac{BR^2 - 2 \cdot BR + 1}{n_{batch}} \tag{2}$$

$$MLin = \frac{BR - 2 \cdot BR + 1}{n_{batch}} \tag{3}$$

where *BR* is the build rate calculated from the predicted process parameters and n_{batch} is the batch size. These terms are weighted by a beta value (<1) and added to the mean squared error of the relative density as presented by Alpaydin [29]. Both types result in an increased mean BR and a slight increase in the prediction error from ~0.11% to ~0.16%. Higher beta values do not increase the build rate further but reach faster an unwanted saturation of the model. The same effect can be observed by using the *MLin* hint.

The differences between both approaches become visible by looking at the predicted process parameters in detail. In the following, process parameters are generated for relative densities from 70% to 99% at a constant layer thickness of 0.05 mm using both models. For both models, the process parameters show an increasing ED at higher relative densities. For the database approach, the noise is higher compared to the ANN, see Figure 11a.

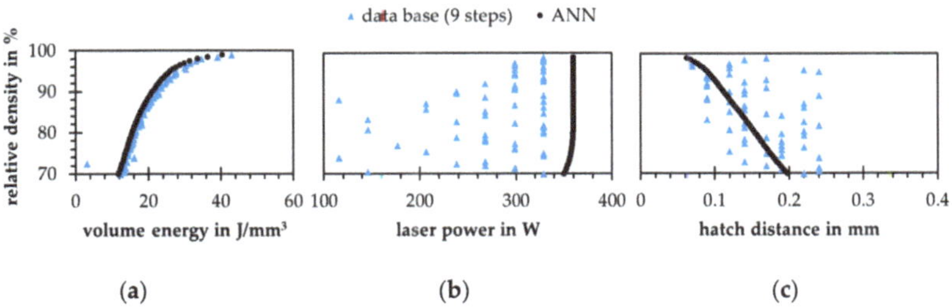

Figure 11. Results of the process parameter prediction for a constant layer thickness of 0.05 mm for both inverse models trained with artificial data, target relative density over (**a**) volume energy; (**b**) laser power; (**c**) hatch distance.

For the process parameters, the database approach shows strong jumps of the process parameters over the relative density, as exemplified by the laser power and the track distance, see Figure 11b,c. The effect can be attributed to single case mapping of the database. These results show that this approach is able to reproduce the origin distribution (ED and relative density), but does not contain information on the concept of process parameter optimization for the present process. As visible in Figure 10c, the process parameters resulting from the database approach reach the correct relative densities as recently shown by Park et al. [13]. Since the resulting densities are predicted by the forward model, there could be a nexus of both models' errors, shown by the jumps of the resulting process parameters. The ANN model as a mean value learner shows here more continuous results. It selects a higher laser power over the entire range of relative density values and a decreasing hatch distance with increasing relative density. Increasing the number of steps in the database approach from 9 to 18 results in a higher average BR which corresponds to that of the ANN model. In addition, the size of the jumps in the individual process parameters is reduced and the model approaches the inverse ANN.

Even with 18 levels and thus more than 100 k combinations (Park et al. [13] used 73 k combinations), the database approach still shows considerable jumps in the results. The reduction in the jumps can be explained by the fact that more possible process parameter combinations are available, resulting in a mean value approximation. Since not every density is exactly represented in the database, a selection must be made according to the target density with soft limits. This limit must be selected the softer, the fewer data are contained in the database. In addition, the calculation time increases with an increasing number of database instances.

Proof of Hypothesis 2. The inverse problem of process parameter prediction can be solved using both a database and a concatenated inverse ANN model. Both models show that they can learn and rebuild the initial distribution related to the ED in a problem-specific manner. In addition, the models can be used to maximize the BR over the entire range of relative densities studied compared to the baseline distribution. For an in-process application, the concatenated ANN shows a higher potential due to the smoothed trajectories as well as an increased tractability of the learned strategy. Thus, the ANN model presents a maximization of the laser power and adjusts the required energy densities via the hatch distance and the scan speed. On the other hand, no generally valid strategy can be determined for the database approach. □

4.5. Real Data Application

After proving the solvability of the inverse problem using synthetic data, the methods are adapted to the real data collected in the process. The data set presented in Section 4.2. is extended to 700 instances considering the process noise with artificial normally distributed noise. Subsequently, it is split into 490 training and 210 test instances. A maximum noise of 5% is assumed for the process parameters. Density noise is assumed to increase linearly towards lower relative densities. The increase in the standard deviation is calculated according to:

$$\sigma_\rho = -0.16 \cdot \rho + 16.16, \quad (4)$$

where ρ represents the original relative density value and σ_ρ represents the assumed standard deviation for the density value. Using the calculated standard deviations, a random normally distributed noise is applied to the original data. Model training shows that considerably more training cycles are required and that an architecture with 8 and 4 neurons in the inner layers leads to highly noisy results. With increasing model complexity, more neurons in the inner layers, the results for the models improve. This is especially evident for the training data. For the test data, there is also a decreasing scatter, which, however, exceeds that of the training data considerably, see Figure 12. Because of the noise of the real data, the mean absolute error loss is used for model training. This loss function is more insensitive against noisy data [39].

Due to the split into training and test data and the factor step-based experimental design, there may be increased gaps that cannot be covered in the training process. Thus, the high dispersion of the predicted densities for the test data may result from the low variance of the data set. A test series with higher variance should lead to improved results, which should be comparable to those of the purely synthetic data. Further improvements could be achievable with the implementation of process and material models (PDE). This way, the model can benefit from data and knowledge combined in a PINN [24,40] also shown for melt pool fluid dynamic prediction [41].

For the inverse model, the forward model from Figure 12c is used. With real data and more neurons, the training times are considerably higher. The result with 40 and 60 neurons on the two inner layers is shown in Figure 13. This setup is chosen because it represents the inverse setup of the forward model. If a smaller number of neurons is chosen for the backward model, the results deteriorate, sometimes significantly. As shown in Figure 13a, the process parameters predicted by the backward model, according to the forward model, reach the specified relative densities. The error influence of the forward model has to be taken into account. The resulting ED of the predicted process parameters is oriented to the minimum of the range given by the experimental data, see Figure 13b.

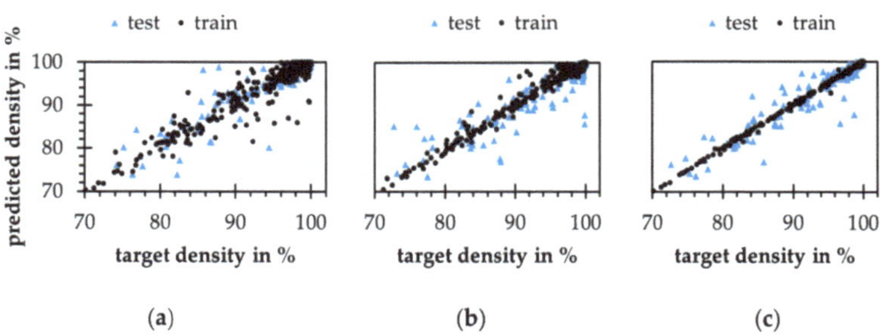

Figure 12. Results of ANN models with the extended experimental data with (**a**) 8 and 4; (**b**) 24 and 12; (**c**) 60 and 40 neurons in the two inner layers.

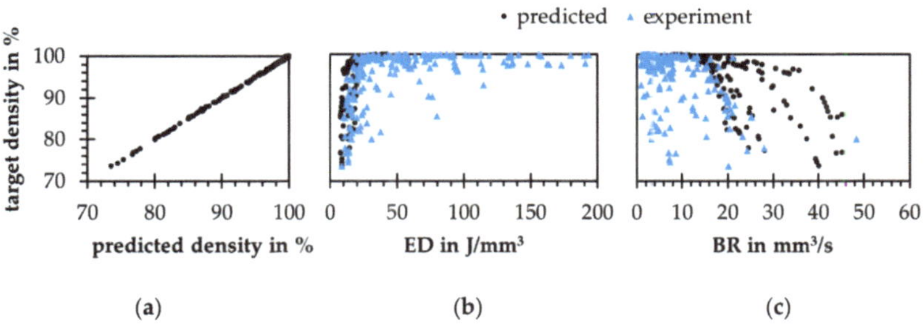

Figure 13. Results of the chained ANN approach with real data, target densities over (**a**) relative densities predicted by the forward model from the process parameters predicted by the backward model; (**b**) ED from the predicted and experimental process parameters; (**c**) build rates of the predicted and experimental process parameters.

If the relative densities from the test data set are applied to the model, as well as the associated layer thickness, the predicted process parameters according to the forward model meet the specifications with a very high accuracy. If the entire considered density range between 70% and 99.5% at a constant layer thickness of 50 µm is used for the prediction of process parameters, a strong model deviation for relative densities below 75% is shown, see Figure 14a. The laser power is chosen almost constantly by the model, slightly decreasing towards higher densities. The hatch distance is lowered considerably at higher relative densities and thus increasing ED is implemented, see Figure 14b,c.

The accuracy of the backward model depends on the accuracy of the forward model. The backward model can never be better than the forward model for the prediction of the relative density. Thus, with the existing experimental data, the model cannot be meaningfully improved and generalized. More data points are needed at this point. The presented approach can be implemented in practice if the parameter space is constrained. It is assumed, but still not proven, that a process parameter set prediction model is machine and material specific. Therefore, a parameter predictor must be derived from previous experimental data collected and processed as described in this paper. It is worth noticing that the presented model is developed on only small specimens. For future works, the influence of part geometry and volume needs to be taken into account. The residual stresses could further be an additional constraint for the inverse model. Next to increasing the build rate, this can improve the process stability decreasing the amount of scrap parts.

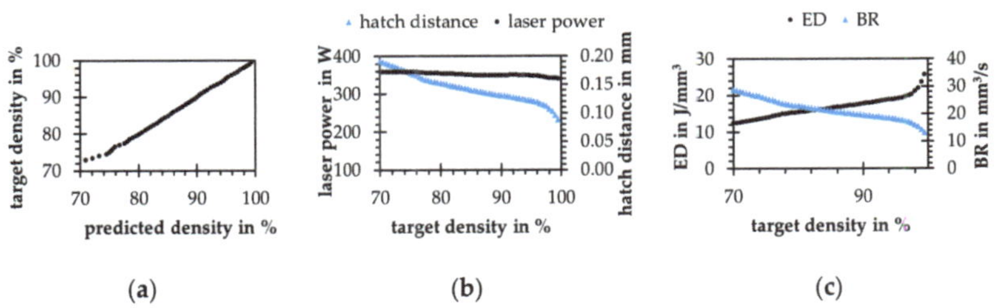

Figure 14. Results of the concatenated ANN approach with the experimental data for given relative densities in 0.5% steps between 70% and 99.5% at a constant layer thickness of 50 µm; (**a**) relative densities predicted by the forward model over the given; (**b**) laser power and hatch distance over the given relative density at a constant scan speed of 3000 mm/s; (**c**) energy density and build rate of the predicted process parameters over the given target density.

4.6. Summary of the Results

1. Boundaries of the process window reached with the statistical test series and different kinds of pores and mechanisms could be mapped.
2. Problems with statistical test series for machine learning are detected and evaluated for the article's target of linking the relative density and the LPBF process parameters.
3. Theoretical solvability of the inverse problem evaluated by synthetic data for both model approaches (concatenated ANN and database).
4. Database approach shows strong jumps for chosen process parameters, while the concatenated ANNs choosing process parameters smooth and strategic
5. By adding hints, the concatenated ANN model is guided by learning to increase the build rates of predicted process parameters.
6. Concatenated ANNs could learn real data problems, but pure quality and fuzziness of the real data worsen the results of the models.

5. Conclusions

It can be stated that both investigated model approaches can be used for solving the inverse problem considered; the prediction of process parameters for given relative densities in the LPBF process. The novelty of this work is the concatenated ANN architecture, which is also trained with real data and provides a technological benefit through process optimization. Despite learning the forward model of a synthetic correlation and although the predicted process parameters follow the ED density correlation model, from an expert point of view, the database approach shows arbitrary jumps in process parameter selection. The ability to reach the target density can be doubted. For the synthetic data, the concatenated ANN model is able to predict the relative density with an R^2 value of 0.98. The backward model is trimmed to higher build rates by the formulation of a hint during the training process. By this, a 12% increase in the mean BR is possible.

The models with real data, which have significantly lower variance, show significantly worse results. Despite artificial noise, the gaps between the data points are too large for a generalized and robust model. Thus, it can be stated that a statistical experimental design is suitable for a process investigation but has too low a data quality for the training of machine learning models. Interpolation between factor levels would be possible, but this results in an almost entirely synthetic dataset similar to the purely synthetic dataset used for model building. Additionally, the results of the models trained with real data are worsened by the noise of the data. While the ED can generally be used to estimate the relative density, a single ED can result in multiple relative density values and higher EDs do not necessarily affect relative densities. Thus, the ED cannot serve as the single indicator for the resulting relative density. Instead, the process–property relationship needs to be understood.

The lack of process parameter predictor models derived from analytical models or simulations enforces a data-driven method. Even physical model-driven trained predictor functions (such as PINNs) rely on specific material and process models, which are not available or accurate enough for an additive metal powder process. Solving the inverse problem of a data-driven model is difficult due to the lack of explainability and tractability of any data-driven model approximation. Most forward models cannot be inverted with any mathematical method. Combining a forward and backward model in a hybrid model that is trained in conjunction can overcome this limitation.

The concatenation of forward and inverse models implies a dependence on the generalization and prediction accuracy of the models. This study shows the theoretical solvability of the inverse problem via concatenated ANNs, but real-world data with higher variance are required to verify the solution. Compared to GANs, the training of the chained models is supervised and related to one feature (here the relative density). By means of a hint, the BR can be increased additionally. Further, there is no need to unfold the whole result space to select a single set of process parameters. Thus, in application, LPBF machine costs for components may be reduced by using a suitable relative density. However, when defining a relative density for specific LPBF components, it is necessary to understand and exploit the interactions between the results of the LPBF process and subsequent process steps, such as heat treatments or hot isostatic pressing, to achieve the desired advantages.

Author Contributions: Conceptualization M.L.A., S.B. and C.W.; methodology, validation, and investigation, M.L.A., S.B. and C.W.; writing—original draft preparation M.L.A. and S.B.; writing—review and editing, S.B., R.F.-H., C.W. and A.T.; visualization, M.L.A. and A.T.; supervision, S.B., C.W., R.F.-H. and A.T. All authors have read and agreed to the published version of the manuscript.

Funding: This research was funded by the European Union from the European Regional Development Fund (ERDF) within the Bremen aerospace research program in Germany, Grant number LURAFO2002A. The author is responsible for the contents of this publication.

Institutional Review Board Statement: Not applicable.

Informed Consent Statement: Not applicable.

Data Availability Statement: Models and data available at: https://doi.org/10.5281/zenodo.7156655.

Conflicts of Interest: The authors declare no conflict of interest.

Abbreviations

ANN	Artificial neural network	MSQ	Mean squared loss hint
BR	Build rate	MVAE	Modified variational autoencoder
D_S	Layer thickness	n_{batch}	Batch size
EBM	Electron beam melting	NSGA-II	Non-dominated sorting genetic algorithm
ED	Energy density	PDE	Partial differential equation
g	Transfer function	PINN	Physics-informed neural network
GAN	Generative adversarial network	P_L	Laser power
HIP	Hot isostatic pressing	ρ_{rel}	Relative density
h_S	Hatch distance	ReLU	Rectified linear unit
INN	Invertible neural network	ROI	Region of interest
LMD	Laser metal deposition	σ_ρ	Standard deviation related to the relative density
LPBF	Laser powder bed fusion, also also powder base fusion with laser beam (PBF-LB) and selective laser melting (SLM)	v_S	Scan speed
MLin	Mean linear loss hint	ω_i	Internal model parameters
MMD	Maximum mean discrepancy	x, y	Input/feature variable

References

1. Vora, J.; Parmar, H.; Chaudhari, R.; Khanna, S.; Doshi, M.; Patel, V.V. Experimental investigations on mechanical properties of multi-layered structure fabricated by GMAW-based WAAM of SS316L. *J. Mater. Res. Technol.* **2022**, *20*, 2748–2757. [CrossRef]
2. Wycisk, E. *Ermüdungseigenschaften der Laseradditiv Gefertigten Titanlegierung TiAl6V4*; Springer: Berlin/Heidelberg, Germany, 2017.
3. Yap, C.Y.; Chua, C.K.; Dong, Z.L.; Liu, Z.H.; Zhang, D.Q.; Loh, L.E.; Sing, S.L. Review of selective laser melting: Materials and applications. *Appl. Phys. Rev.* **2015**, *2*, 041101. [CrossRef]
4. Herzog, D.; Seyda, V.; Wycisk, E.; Emmelmann, C. Additive manufacturing of metals. *Acta Mater.* **2016**, *117*, 371–392. [CrossRef]
5. Thomas, D.S.; Gilbert, S.W. *Costs and Cost Effectiveness of Additive Manufacturing*; National Institute of Standards and Technology: Gaithersburg, MD, USA, 2014.
6. Ma, M.; Wang, Z.; Gao, M.; Zeng, X. Layer thickness dependence of performance in high-power selective laser melting of 1Cr18Ni9Ti stainless steel. *J. Mater. Process. Technol.* **2015**, *215*, 142–150. [CrossRef]
7. Bertoli, U.S.; Wolfer, A.J.; Matthews, M.J.; Delplanque, J.P.R.; Schoenung, J.M. On the limitations of volumetric energy density as a design parameter for selective laser melting. *Mater. Des.* **2017**, *113*, 331–340. [CrossRef]
8. Liu, W.; Chen, C.; Shuai, S.; Zhao, R.; Liu, L.; Wang, X.; Hu, T.; Xuan, W.; Li, C.; Yu, J.; et al. Study of pore defect and mechanical properties in selective laser melted Ti6Al4V alloy based on X-ray computed tomography. *Mater. Sci. Eng. A* **2020**, *797*, 139981. [CrossRef]

9. Gebhardt, A. *Generative Fertigungsverfahren: Rapid Prototyping-Rapid Tooling-Rapid Manufacturing*, 3rd ed.; Hanser: München Germany, 2007; ISBN 978-3-446-22666-1.
10. Herzog, D.; Bartsch, K.; Bossen, B. Productivity optimization of laser powder bed fusion by hot isostatic pressing. *Addit. Manuf* **2020**, *36*, 101494. [CrossRef]
11. Vasileska, E.; Demir, A.G.; Colosimo, B.M.; Previtali, B. A novel paradigm for feedback control in LPBF: Layer-wise correction for overhang structures. *Adv. Manuf.* **2022**, *10*, 326–344. [CrossRef]
12. Liverani, E.; Fortunato, A. Additive manufacturing of AISI 420 stainless steel: Process validation, defect analysis and mechanical characterization in different process and post-process conditions. *Int. J. Adv. Manuf. Technol.* **2021**, *117*, 809–821. [CrossRef]
13. Park, H.S.; Nguyen, D.S.; Le-Hong, T.; Van Tran, X. Machine learning-based optimization of process parameters in selective laser melting for biomedical applications. *J. Intell. Manuf.* **2022**, *33*, 1843–1858. [CrossRef]
14. Bai, S.; Perevoshchikova, N.; Sha, Y.; Wu, X. The effects of selective laser melting process parameters on relative density of the AlSi10Mg parts and suitable procedures of the archimedes method. *Appl. Sci.* **2019**, *9*, 583. [CrossRef]
15. Shamsdini, S.; Shakerin, S.; Hadadzadeh, A.; Amirkhiz, B.S.; Mohammadi, M. A trade-off between powder layer thickness and mechanical properties in additively manufactured maraging steels. *Mater. Sci. Eng. A* **2020**, *776*, 139041. [CrossRef]
16. Welsch, A.; Eitle, V.; Buxmann, P. Maschinelles Lernen. *HMD Prax. Wirtsch.* **2018**, *55*, 366–382. [CrossRef]
17. Minbashian, A.; Bright, J.E.; Bird, K.D. A comparison of artificial neural networks and multiple regression in the context of research on personality and work performance. *Organ. Res. Methods* **2010**, *13*, 540–561. [CrossRef]
18. Lee, J.W.; Park, W.B.; Do Lee, B.; Kim, S.; Goo, N.H.; Sohn, K.S. Dirty engineering data-driven inverse prediction machine learning model. *Sci. Rep.* **2020**, *10*, 20443. [CrossRef]
19. Yu, X.; Chu, Y.; Jiang, F.; Guo, Y.; Gong, D. SVMs classification based two-side cross domain collaborative filtering by inferring intrinsic user and item features. *Knowl.-Based Syst.* **2018**, *141*, 80–91. [CrossRef]
20. Yu, X.; Yang, J.; Xie, Z. Training SVMs on a bound vectors set based on Fisher projection. *Front. Comput. Sci.* **2014**, *8*, 793–806 [CrossRef]
21. Ardizzone, L.; Kruse, J.; Wirkert, S.; Rahner, D.; Pellegrini, E.W.; Klessen, R.S.; Maier-Hein, L.; Rother, C.; Köthe, U. Analyzing inverse problems with invertible neural networks. *arXiv* **2019**, arXiv:1808.04730. [CrossRef]
22. Creswell, A.; White, T.; Dumoulin, V.; Arulkumaran, K.; Sengupta, B.; Bharath, A.A. Generative adversarial networks: An overview. *IEEE Signal Process. Mag.* **2018**, *35*, 53–65. [CrossRef]
23. Mohammad-Djafari, A. Regularization, Bayesian inference, and machine learning methods for inverse problems. *Entropy* **2021**, *23*, 1673. [CrossRef] [PubMed]
24. Karniadakis, G.E.; Kevrekidis, I.G.; Lu, L.; Perdikaris, P.; Wang, S.; Yang, L. Physics-informed machine learning. *Nat. Rev. Phys.* **2021**, *3*, 422–440. [CrossRef]
25. Wang, J.; Wang, Y.; Chen, Y. Inverse Design of Materials by Machine Learning. *Materials* **2022**, *15*, 1811. [CrossRef]
26. Kadeethum, T.; Jørgensen, T.M.; Nick, H.M. Physics-Informed Neural Networks for Solving Inverse Problems of Nonlinear Biots Equations: Batch Training. In Proceedings of the 54th U.S. Rock Mechanics/Geomechanics Symposium, Golden, Colorado, USA, 28 June–1 July 2020. physical event cancelled.
27. Raissi, M.; Perdikaris, P.; Karniadakis, G.E. Physics-informed neural networks: A deep learning framework for solving forward and inverse problems involving nonlinear partial differential equations. *J. Comput. Phys.* **2019**, *378*, 686–707. [CrossRef]
28. Alpaydin, E. *Maschinelles Lernen*, 2nd ed.; De Gruyter Oldenbourg: Berlin, Germany; Boston, MA, USA, 2019.
29. Dinh, L.; Sohl-Dickstein, J.; Bengio, S. Density estimation using real nvp. *arXiv* **2016**, arXiv:1605.08803. [CrossRef]
30. Kanne-Schludde, T. Erzeugung Synthetischer Trainingsdaten Für Die Deep Learning Basierte Bestimmung von GPS-Koordinaten Aus Fotos Am Beispiel Der Notre Dame. 2020. Available online: https://users.informatik.haw-hamburg.de/~{}ubicomp/projekte/master2020-proj/kanne_hp.pdf (accessed on 7 October 2022).
31. Batista, G.E.; Monard, M.C. An analysis of four missing data treatment methods for supervised learning. *Appl. Artif. Intell.* **2003**, *17*, 519–533. [CrossRef]
32. Kelleher, J.D.; Mac Namee, B.; D'arcy, A. *Fundamentals of Machine Learning for Predictive Data Analytics: Algorithms, Worked Examples, and Case Studies*; MIT Press: Cambridge, MA, USA, 2020.
33. Brownlee, J. *Data Preparation for Machine Learning: Data Cleaning, Feature Selection, and Data Transforms in Python*; Machine Learning Mastery: San Juan, Puerto Rico, 2020.
34. Facebook, PyTorch: Machine Learning Framework: Facebook. 2021. Available online: https://pytorch.org/ (accessed on 7 October 2022).
35. Liu, Z.; Zhou, J. Introduction to Graph Neural Networks. In *Synthesis Lectures on Artificial Intelligence and Machine Learning*; Springer: Berlin/Heidelberg, Germany, 2020; Volume 14, pp. 1–127. [CrossRef]
36. Ning, J.; Wang, W.; Zamorano, B.; Liang, S.Y. Analytical modeling of lack-of-fusion porosity in metal additive manufacturing. *Appl. Phys. A* **2019**, *125*, 797. [CrossRef]
37. Wang, T.; Dai, S.; Liao, H.; Zhu, H. Pores and the formation mechanisms of SLMed AlSi10Mg. *Rapid Prototyp. J.* **2020**, *26*, 1657–1664. [CrossRef]
38. Kingma, D.P.; Ba, J. Adam: A method for stochastic optimization. *arXiv* **2014**, arXiv:1412.6980. [CrossRef]

39. Murphy, K.P. *Machine Learning: A Probabilistic Perspective*; MIT Press: Cambridge, MA, USA; London, UK, 2012.
40. Cai, S.; Mao, Z.; Wang, Z.; Yin, M.; Karniadakis, G.E. Physics-informed neural networks (PINNs) for fluid mechanics: A review. *Acta Mech. Sin.* **2021**, *37*, 1727–1738. [CrossRef]
41. Zhu, Q.; Liu, Z.; Yan, J. Machine learning for metal additive manufacturing: Predicting temperature and melt pool fluid dynamics using physics-informed neural networks. *Comput. Mech.* **2021**, *67*, 619–635. [CrossRef]

MDPI AG
Grosspeteranlage 5
4052 Basel
Switzerland
Tel.: +41 61 683 77 34

Materials Editorial Office
E-mail: materials@mdpi.com
www.mdpi.com/journal/materials

www.ingramcontent.com/pod-product-compliance
Lightning Source LLC
Chambersburg PA
CBHW041344210926
51850CB00013B/2983